CW01023336

MATRIX THEORY & APPLICATIONS
FOR SCIENTISTS & ENGINEERS

Alexander Graham, M.A., M.Sc., Ph.D.

Dover Publications, Inc.
Mineola, New York

Copyright

Copyright © 1979, 2018 by Alexander Graham
All rights reserved.

Bibliographical Note

This Dover edition, first published in 2018, is an unabridged and corrected republication of *Matrix Theory and Applications for Engineers and Mathematicians,* originally published in 1979 by Ellis Horwood, Ltd, Chichester, West Sussex.

International Standard Book Number

ISBN-13: 978-0-486-82419-2
ISBN-10: 0-486-82419-5

Manufactured in the United States by LSC Communications
82419501 2018
www.doverpublications.com

Table of Contents

EXAMPLES OF NOTATION USED

$a_{ij}, b_{ij}, \ldots a_{mn}$	elements of a matrix
$\alpha, \beta, \ldots$	scalars
$A, B, \ldots$	matrices
I	the unit matrix
O	the zero matrix
A'	the transpose of A
$\bar{A}$	the conjugate of A
$\bar{A}'$ or A^*	the conjugate transpose of A
$\mathbf{a}, \mathbf{x}, \ldots$	vectors
$\mathbf{A}_i, \mathbf{B}_i, \ldots$	i^{th} column (i.e. vector) of $A, B, \ldots$
$\bar{\mathbf{x}}$	the conjugate of $\mathbf{x}$
$\in$	'belongs to'
$\notin$	'does not belong to'
$\exists$	'there exists'

Preface

I am aware that there are in existence many excellent books on matrix algebra, but it seems to me that very few of them are written with the average student in mind. Some of them are excellent exercises in mathematical rigour, but hardly suitable for a scientist who wishes to understand a particular technique of linear algebra. Others are a summary of various techniques, with little or no attempt at a justification of the underlying principles.

This book steers a middle course between the two extremes. With the rapid use of even more sophisticated techniques in science, economics and engineering, and especially in control theory, it has become essential to have a real understanding of the many methods being used. To achieve simplicity combined with as deep an understanding as possible, I have tried to use very simple notation, even at the possible expense of rigour. I know that a student who, for example, is not used to Gothic lettering, will actually find a passage of mathematics using such symbols (to denote sets) much more difficult to absorb than when Roman lettering is used.

Engineers generally denote vectors by $\mathbf{x}, \mathbf{y}, \ldots$ whereas methematicians will use these letters, but will also use α, $\beta, \ldots$ and other symbols. There is good reason for this; after all, vectors are elements of a set which satisfy the axioms of a vector space. Nevertheless I have followed the engineers' usage for the notation, but have been careful to stress the general nature of a vector.

A few brief remarks about the organization of this book. In Chapter 1 matrices and matrix operations are defined. Chapter 2 deals with vector spaces and the transition matrix. In Chapter 3, various concepts introduced in the first two chapters are combined in the discussion of Linear Transformations. Chapters 4 and 5 deal with various important concepts in matrix theory, such as the rank of a matrix, introduce determinants and solution of sets of simultaneous equations.

Chapters 6 and 7 are, from the engineer's point of view, the most important in the book. Various aspects of eigenvalue and eigenvector concepts are discussed. These concepts are applied to obtain the decomposition of the vector space and to choose a basis relative to which a linear transformation has a block-diagonal matrix representation. Various matrix block-diagonal forms are discussed.

Finally in Chapter 8 techniques for inverting a nonsingular matrix are discussed. I have found some of these techniques very useful especially for various aspects of control theory.

Limitation of length have prevented the inclusion of all aspects of modern matrix theory, but the most important have been covered.

Thus, there is enough material to make this book a useful one for college and university students of mathematics, engineering, and science who are anxious not only to learn but to understand.

A. GRAHAM

CHAPTER 1

Matrices

1.1 Matrices and Matrix Operations

Definition 1.1
A matrix is a rectangular array of numbers of the form

$$A = \begin{bmatrix} a_{11} & a_{12} & \cdots & a_{1n} \\ a_{21} & a_{22} & \cdots & a_{2n} \\ \vdots & & & \\ a_{m1} & a_{m2} & \cdots & a_{mn} \end{bmatrix}$$

In the above form the matrix has m rows and n columns, we say that it is of **order** $m \times n$.

The mn numbers $a_{11}, a_{12}, \ldots, a_{mn}$ are known as the **elements** of the matrix A. If the elements belong to a field F, we say that A is a matrix **over** the field F. Since the concept of a field plays an emportant role in development of vector spaces we shall state the axioms for a field at this early stage.

The notation we use is the following

$a \in S$ means that a belongs to the set S

$a \notin S$ means that a does not belong to the set S

ϕ denotes the empty set

$\exists$ denotes 'there exists'.

A field F consists of a non-empty set of elements and two laws called **addition** and **multiplication** for combining elements satisfying the following axioms:

Let $a,b,c \in F$

$A1$ $a + b$ is a unique element $\in F$

$A2$ $(a+b) + c = a + (b+c)$

$A3$ $\exists 0 \in F$ such that $0 + a = a$, all $a \in F$.
$A4$ For each $a \in F$, $\exists (-a) \in F$ such that $a + (-a) = 0$
$A5$ $a + b = b + a$
$M1$ ab is a unique element $\in F$
$M2$ $(ab)c = a(bc)$
$M3$ $\exists 1 \in F$ such that $1a = a$ for all $a \in F$
$M4$ If $a \neq 0$, $\exists a^{-1} \in F$ such that $a^{-1}a = 1$.
$M5$ $ab = ba$
$M6$ $a(b+c) = ab + ac$.

Typical examples of fields are: real numbers, rational numbers, and complex numbers.

If we consider the element a_{ij} of a matrix the first suffix, i, indicates that the element stands in the ith row, whereas the second suffix, j, indicates that it stands in the jth column.

For example, the element which stands in the second row and the first column is a_{21}.

We usually denote a matrix by a capital letter, say A, or by its $(i,j)^{\text{th}}$ element in the form $[a_{ij}]$.

Definition 1.2
A **square matrix** is a matrix of order $n \times n$ of the form

$$A = \begin{bmatrix} a_{11} & a_{12} & \dots & a_{1n} \\ a_{21} & a_{22} & \dots & a_{2n} \\ \vdots & & & \\ a_{n1} & a_{n2} & \dots & a_{nn} \end{bmatrix}$$

The elements $a_{11}, a_{22}, \dots, a_{nn}$ of A are called **diagonal elements.**

Example 1.1
The following are matrices:

$$A = \begin{bmatrix} 1 & 0 & 2 \\ 2 & -1 & 2 \\ 0 & 0 & 1 \end{bmatrix}, \quad B = \begin{bmatrix} 1 & 2 & 3 \\ 3 & 0 & -2 \end{bmatrix}, \quad C = \begin{bmatrix} 1 \\ 1 \\ 0 \end{bmatrix} \quad D = [2,1].$$

A is a square matrix of order 3×3.
B is a rectangular matrix of order 2×3.
C and D are matrices of order 3×1 and 1×2 respectively.

Note
C is also known as a **column** matrix or **column vector**. D is known as a **row** matrix or a **row vector**.
The diagonal elements of A are $1, -1, 1$.
The $(2, 3)$ element of B is $b_{23} = -2$.

Definitions 1.3
(1) The **zero** matrix of order $m \times n$ is the matrix having all its mn elements equal to zero.
(2) The **unit** or **identity** matrix I is the square matrix of order $n \times n$ whose diagonal elements are all equal to 1 and all remaining elements are 0.
(3) The **diagonal matrix** A is a square matrix for which $a_{ij} = 0$ whenever $i \neq j$. Thus all the off-diagonal elements of a diagonal matrix are zero.
(4) The diagonal matrix for which all the diagonal elements are equal to each other is called a **scalar matrix**.

Example 1.2
Consider the following matrices:

$$A = \begin{bmatrix} 1.5 & 0 & 0 \\ 0 & 1.5 & 0 \\ 0 & 0 & 1.5 \end{bmatrix}, \quad B = \begin{bmatrix} 2 & 0 & 0 \\ 0 & -3 & 0 \\ 0 & 0 & -1 \end{bmatrix}$$

$$I = \begin{bmatrix} 1 & 0 & 0 \\ 0 & 1 & 0 \\ 0 & 0 & 1 \end{bmatrix}, \quad O = \begin{bmatrix} 0 & 0 & 0 \\ 0 & 0 & 0 \end{bmatrix}$$

A is a scalar matrix of order 3×3
B is a diagonal matrix of order 3×3
I is the unit matrix of order 3×3
O is the zero matrix of order 2×3.

Note
We shall use the accepted convention of denoting the zero and unit matrices by O and I respectively. It is of course necessary to state the order of matrices considered unless this is obvious from the text.

Definition 1.4

(1) An **upper triangular** matrix A is a square matrix whose elements $a_{ij} = 0$ for $i > j$.

(2) A **lower triangular** matrix A is a square matrix whose elements $a_{ij} = 0$ for $i < j$.

Example 1.3

Consider the two matrices

$$A = \begin{bmatrix} 1 & 2 & 0 \\ 0 & -3 & 1 \\ 0 & 0 & 5 \end{bmatrix} \text{ and } B = \begin{bmatrix} 1 & 0 & 0 \\ 2 & 2 & 0 \\ 0 & -1 & 3 \end{bmatrix}$$

A is an upper triangular matrix
B is a lower triangular matrix.

Definition 1.5

Two matrices $A = [a_{ij}]$ and $B = [b_{ij}]$ are said to be **equal** if

(1) A and B are of the same order, and

(2) the corresponding elements are equal, that is if $a_{ij} = b_{ij}$ (all i and j).

Operations on Matrices

Definition 1.6

The **sum** (or difference) of two matrices $A = [a_{ij}]$ and $B = [b_{ij}]$ of the same order, say $m \times n$, is a matrix $C = [c_{ij}]$ also of order $m \times n$ such that

$$c_{ij} = a_{ij} + b_{ij} \ (i = 1, \ldots m; j = 1, \ldots n).$$

(Or $c_{ij} = a_{ij} - b_{ij}$ if we are considering the difference of A and B.)

Example 1.4

Given

$$A = \begin{bmatrix} 1 & 2 & -3 \\ 0 & 1 & -1 \end{bmatrix} \text{ and } B = \begin{bmatrix} -1 & 1 & 2 \\ -2 & 0 & 1 \end{bmatrix}$$

find $A+B$ and $A-B$.

Solution

$$A + B = \begin{bmatrix} 0 & 3 & -1 \\ -2 & 1 & 0 \end{bmatrix} \text{ and } A - B = \begin{bmatrix} 2 & 1 & -5 \\ 2 & 1 & -2 \end{bmatrix}$$

Definition 1.7
The multiplication of a matrix $A = [a_{ij}]$ of order $m \times n$ by a scalar r is the matrix rA such that

$$rA = [ra_{ij}] = r[a_{ij}] \quad (r = 1, \ldots m; j = 1, \ldots n)$$

Example 1.5
Given

$$A = \begin{bmatrix} 1 & 2 & -3 \\ 0 & 1 & -1 \end{bmatrix}, \text{ find } 3A.$$

Solution

$$3A = \begin{bmatrix} 3 & 6 & -9 \\ 0 & 3 & -3 \end{bmatrix}$$

Note that $3A = A + A + A$, and we can evaluate the right-hand side by definition 1.6.

Definition 1.8
The **product** of the matrix $A = [a_{ij}]$ of order $m \times l$ and the matrix $B = [b_{ij}]$ of order $l \times n$ is the matrix $C = [c_{ij}]$ of order $m \times n$ defined by

$$\begin{aligned} c_{ij} &= \sum_{r=1}^{l} a_{ir} b_{rj} & i = 1,2, \ldots m \\ &= a_{i1} b_{1j} + a_{i2} b_{2j} + \ldots + a_{il} b_{lj} & j = 1,2, \ldots n \ . \end{aligned}$$

We illustrate this definition by showing up the elements of A and B making up the $(i,j)^{\text{th}}$ element of C.

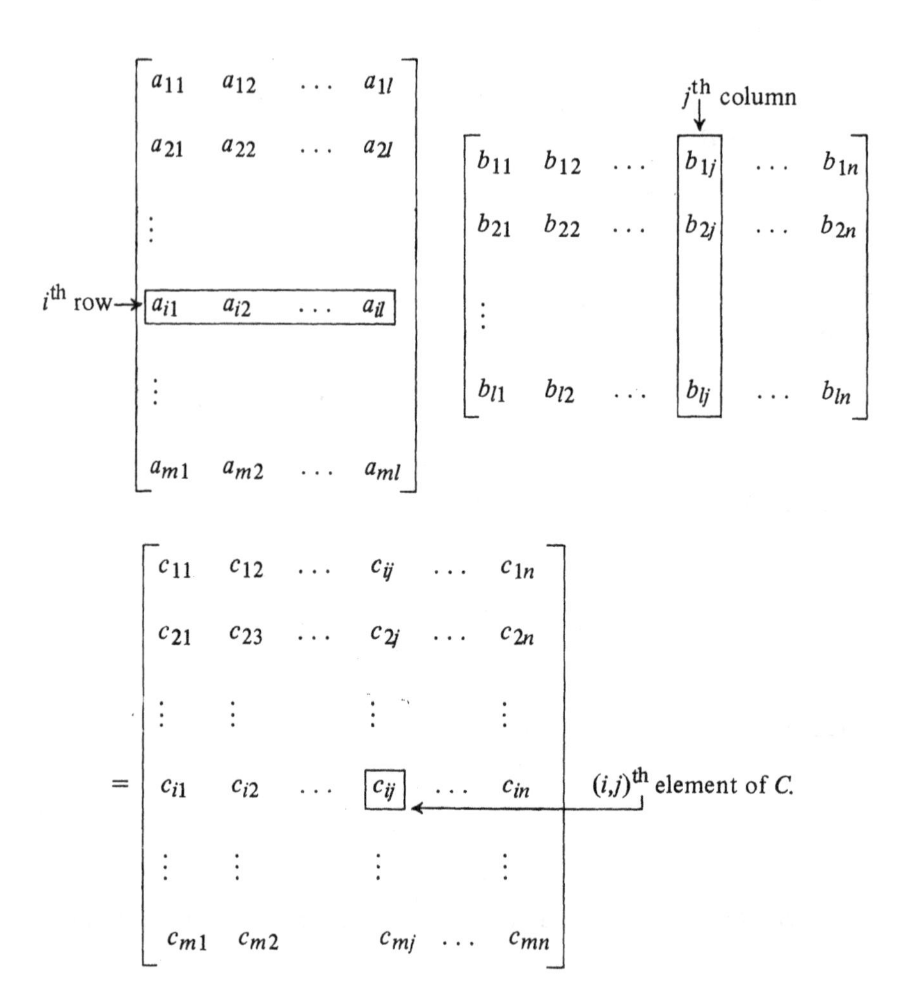

Note the following:

(i) The product AB is defined only if the number of columns of A is the same as the number of rows of B. If this is the case, we say that A is **conformable** to B.

(ii) The $(i,j)^{\text{th}}$ element of C is evaluated by using the ith row of A and j^{th} column of B.

(iii) If A is conformable to B the product AB is defined, but it does not follow that the product BA is defined. Indeed, if A and B are of order $m \times l$ and $l \times n$ respectively, the product AB is defined but the product BA is not, unless $n = m$.

(iv) When both AB and BA are defined, $AB \neq BA$ in general, that is, matrix multiplication is NOT commutative.

(v) (a) If $AX = 0$, it does not necessarily follow that $A = 0$ or $X = 0$.
(b) If $AB = AC$, it does not necessarily follow that $B = C$.
(See Sec. 1.2 for further discussion of (iv) and (v) and examples).

Example 1.6

$$A = \begin{bmatrix} 1 & 0 & -2 \\ 2 & 1 & -1 \end{bmatrix}, \quad B = \begin{bmatrix} 2 & 0 & 2 \\ -1 & 1 & -2 \\ 1 & 0 & 1 \end{bmatrix}, \quad C = \begin{bmatrix} 1 \\ 2 \\ 3 \end{bmatrix}$$

Find (if possible)
(i) AB, (ii) BA, (iii) AC, (iv) CA, (v) BC.

Solution

(i) Since A has 3 columns and B has 3 rows, the product AB is defined.

$$AB = \begin{bmatrix} 0 & 0 & 0 \\ 2 & 1 & 1 \end{bmatrix}.$$

(ii) Since B has 3 columns and A has 2 rows the product BA is not defined.

(iii) Since A has 3 columns and C has 3 rows, AC is defined

$$AC = \begin{bmatrix} -5 \\ 1 \end{bmatrix}.$$

(iv) The product CA is not defined.

(v)

$$BC = \begin{bmatrix} 8 \\ -5 \\ 4 \end{bmatrix}.$$

Example 1.7
Given the matrices:

$$A = \begin{bmatrix} 1 & -2 & 3 \\ 0 & 1 & -1 \\ 3 & -3 & 0 \end{bmatrix}, \qquad X = \begin{bmatrix} x_1 \\ x_2 \\ x_3 \end{bmatrix}, \qquad B = \begin{bmatrix} 1 \\ 2 \\ 3 \end{bmatrix}$$

write the equations $AX = B$ in full.

Solution
$AX = B$ is the equation:

$$\begin{bmatrix} 1 & -2 & 3 \\ 0 & 1 & -1 \\ 3 & -3 & 0 \end{bmatrix} \begin{bmatrix} x_1 \\ x_2 \\ x_3 \end{bmatrix} = \begin{bmatrix} 1 \\ 2 \\ 3 \end{bmatrix}.$$

By Def. 1.8, we can write the above as

$$\begin{bmatrix} x_1 - 2x_2 + 3x_3 \\ 0 + x_2 - x_3 \\ 3x_1 - 3x_2 + 0 \end{bmatrix} = \begin{bmatrix} 1 \\ 2 \\ 3 \end{bmatrix}.$$

By Def. 1.5 the above equation is equivalent to the following system of simultaneous equations:

$$\begin{aligned} x_1 - 2x_2 + 3x_3 &= 1 \\ x_2 - x_3 &= 2 \\ 3x_1 - 3x_2 &= 3. \end{aligned}$$

Definition 1.9
The **transpose** of a matrix $A = [a_{ij}]$ of order $m \times n$ is the matrix $A' = [b_{ij}]$ of order $n \times m$ obtained from A by interchanging the rows and columns of A so that $b_{ij} = a_{ji}$ $(i = 1,2, \ldots n, j = 1,2, \ldots m)$, for example if

$$A = \begin{bmatrix} 1 & -2 \\ 0 & 1 \\ -1 & 2 \end{bmatrix}, \qquad \text{then } A' = \begin{bmatrix} 1 & 0 & -1 \\ -2 & 1 & 2 \end{bmatrix},$$

The transpose of a column matrix is a row matrix and vice versa, thus if

$$A = \begin{bmatrix} 1 \\ 2 \\ 3 \end{bmatrix} \text{ then } A' = [1 \quad 2 \quad 3].$$

Note that $(A')' = A$.

Notation

To denote vectors (see Def. 2.1) we shall make use of several notations.

When considering a one-column matrix or a one-row matrix, we use

$$\begin{bmatrix} 1 \\ 2 \\ 3 \end{bmatrix} \text{ and } [1,2,3] \text{ respectively.}$$

Since a one-row matrix is less space-consuming to write than a one-column matrix, we frequently write the column matrix in the transposed form as $[1,2,3]'$.

Finally, if it is immaterial whether the vectors under discussion are row or column vectors, we use the notation (1,2,3).

1.2 SOME PROPERTIES OF MATRIX OPERATIONS

In this section we shall state without proof (in general) a number of properties of matrix operation. The interested reader will find the proofs in most of the books mentioned in the Bibliography.

Although there are a number of analogies between the algebraic properties of matrices and real numbers, there are also striking differences, some of which will be pointed out.

Let A, B, and C be matrices, each of order $m \times n$

Addition laws

(1) Matrix addition is commutative; that is,

$$A + B = B + A.$$

(2) Matrix addition is associative; that is,

$$A + (B + C) = (A + B) + C.$$

(3) There exists a unique zero matrix O of order $m \times n$ such that

$$A + O = O + A = A.$$

(4) Given the matrix A, there exists a unique matrix B such that

$$A + B = 0$$

The above property serves to introduce the **subtraction** of matrices. Indeed the unique matrix B in the above equation is found to be equal to $(-1)A$ which we write as $-A$. We then find that $A - A = 0$ and that $-(-A) = A$.

(5) Multiplication by scalars.
If **r** and **s** are scalars, then
(a) $r(A+B) = rA + rB$
(b) $(r+s)A = rA + sA$
(c) $r(A-B) = rA - rB$
(d) $(r-s)A = rA - sA.$

Example 1.8
Given

$$A = \begin{bmatrix} 1 & -1 & 2 \\ 0 & -2 & 3 \end{bmatrix}, \qquad B = \begin{bmatrix} -2 & 1 & 1 \\ 1 & 0 & 2 \end{bmatrix}$$

and

$$C = \begin{bmatrix} 1 & -1 & 1 \\ 0 & 1 & 0 \end{bmatrix}$$

(a) show that (i) $A + B = B + A$
(ii) $A + (B+C) = (A + B) + C.$

(b) find the matrix X such that $A + X = 0.$

Solution

(a) (i)
$$A + B = \begin{bmatrix} -1 & 0 & 3 \\ 1 & -2 & 5 \end{bmatrix} = B + A.$$

(ii)
$$B + C = \begin{bmatrix} -1 & 0 & 2 \\ 1 & 1 & 2 \end{bmatrix}, \qquad \text{so that}$$

$$A + (B+C) = \begin{bmatrix} 1 & -1 & 2 \\ 0 & -2 & 3 \end{bmatrix} + \begin{bmatrix} -1 & 0 & 2 \\ 1 & 1 & 2 \end{bmatrix} = \begin{bmatrix} 0 & -1 & 4 \\ 1 & -1 & 5 \end{bmatrix}.$$

Also, using result (i) above,

$$(A+B) + C = \begin{bmatrix} -1 & 0 & 3 \\ 1 & -2 & 5 \end{bmatrix} + \begin{bmatrix} 1 & -1 & 1 \\ 0 & 1 & 0 \end{bmatrix} = \begin{bmatrix} 0 & -1 & 4 \\ 1 & -1 & 5 \end{bmatrix}.$$

(b)
$$X = (-1)A = \begin{bmatrix} -1 & 1 & -2 \\ 0 & 2 & -3 \end{bmatrix}.$$

Multiplication laws

(6) If A is $m \times l$ and $B = l \times m$, both AB and BA are defined, but matrix multiplication is not commutative; that is, in general $AB \neq BA$. Note that this is an important difference between the multiplication of matrices and real numbers. If $AB = BA$, we say that A and B **commute** or are **commutative**.

Example 1.9

Given

$$A = \begin{bmatrix} 1 & -1 \\ 2 & -3 \end{bmatrix}, \quad B = \begin{bmatrix} 1 & 2 \\ -2 & 0 \end{bmatrix}, \quad C = \begin{bmatrix} 1 & 2 \\ 2 & 1 \end{bmatrix} \text{ and } D = \begin{bmatrix} -1 & 3 \\ 3 & -1 \end{bmatrix}$$

find AB, BA, CD and DC.

Solution

$$AB = \begin{bmatrix} 1 & -1 \\ 2 & -3 \end{bmatrix} \begin{bmatrix} 1 & 2 \\ -2 & 0 \end{bmatrix} = \begin{bmatrix} 3 & 2 \\ 8 & 4 \end{bmatrix}$$

$$BA = \begin{bmatrix} 1 & 2 \\ -2 & 0 \end{bmatrix} \begin{bmatrix} 1 & -1 \\ 2 & 3 \end{bmatrix} = \begin{bmatrix} 5 & 5 \\ -2 & 2 \end{bmatrix}$$

hence $AB \neq BA$.

$$CD = \begin{bmatrix} 1 & 2 \\ 2 & 1 \end{bmatrix} \begin{bmatrix} -1 & 3 \\ 3 & -1 \end{bmatrix} = \begin{bmatrix} 5 & 1 \\ 1 & 5 \end{bmatrix}$$

$$DC = \begin{bmatrix} -1 & 3 \\ 3 & -1 \end{bmatrix} \begin{bmatrix} 1 & 2 \\ 2 & 1 \end{bmatrix} = \begin{bmatrix} 5 & 1 \\ 1 & 5 \end{bmatrix}$$

hence $CD = DC$.

Example 1.10
Given

$$A = \begin{bmatrix} 3 & -1 \\ -6 & 2 \\ -3 & 1 \end{bmatrix} \quad X = \begin{bmatrix} 1 & 2 \\ 3 & 6 \end{bmatrix}$$

$$B = \begin{bmatrix} 3 & 1 \\ 2 & 2 \end{bmatrix} \text{ and } C = \begin{bmatrix} 2 & -1 \\ -1 & -4 \end{bmatrix}$$

(i) evaluate AX, and

(ii) verify that $AB = AC$.

Comment on the results.

Solution

(i)

$$AX = \begin{bmatrix} 3 & -1 \\ -6 & 2 \\ -3 & 1 \end{bmatrix} \begin{bmatrix} 1 & 2 \\ 3 & 6 \end{bmatrix} = \begin{bmatrix} 0 & 0 \\ 0 & 0 \\ 0 & 0 \end{bmatrix} = [0] .$$

(ii)

$$AB = \begin{bmatrix} 3 & -1 \\ -6 & 2 \\ -3 & 1 \end{bmatrix} \begin{bmatrix} 3 & 1 \\ 2 & 2 \end{bmatrix} = \begin{bmatrix} 7 & 1 \\ -14 & -2 \\ -7 & -1 \end{bmatrix} .$$

$$AC = \begin{bmatrix} 3 & -1 \\ -6 & 2 \\ -3 & 1 \end{bmatrix} \begin{bmatrix} 2 & -1 \\ -1 & -4 \end{bmatrix} = \begin{bmatrix} 7 & 1 \\ -14 & -2 \\ -7 & -1 \end{bmatrix} .$$

In the above example it is shown that
(i) $AX = 0$ does not necessarily imply that $A = 0$ or $X = 0$, and
(ii) $AB = AC$ does not necessarily imply that $B = C$.
These are just two more of the striking differences between operations with matrices and real numbers.

(7) Matrix multiplication is
  (i) distributive with respect to addition; that is,

$$A(B+C) = AB + AC, \text{ and}$$

$$(B+C)A = BA + CA.$$

  provided that the matrices are conformable.

  (ii) associative, that is,

$$A(BC) = (AB)C = ABC$$

  provided that expressions on either side of the equations are defined

Definition 1.10
The rth **power** of a square matrix A of order $n \times n$ is denoted by A^r and defined by

$$A^r = A^{r-1}A \text{ for } r \geq 1$$

and

$$A^0 = I \text{ (I being the unit matrix of order } n \times n).$$

Thus $A^2 = AA$
and $A^3 = AAA$.
The powers of a matrix are commutative, that is

$$A^rA^s = A^{r+s} = A^sA^r \quad (r,s \geq 0).$$

Example 1.11
Given

$$A = \begin{bmatrix} 0 & 1 & 0 \\ 0 & 0 & 1 \\ -4 & -6 & -4 \end{bmatrix}$$

evaluate the matrix polynomial

$$f(A) = A^3 + 4A^2 + 6A + 4I.$$

Solution
We find that

$$A^2 = \begin{bmatrix} 0 & 0 & 1 \\ -4 & -6 & -4 \\ -16 & 20 & 10 \end{bmatrix}$$

and

$$A^3 = \begin{bmatrix} 0 & 1 & 0 \\ 0 & 0 & 1 \\ -4 & -6 & -4 \end{bmatrix} \begin{bmatrix} 0 & 0 & 1 \\ -4 & -6 & -4 \\ 16 & 20 & 10 \end{bmatrix} = \begin{bmatrix} -4 & -6 & -4 \\ 16 & 20 & 10 \\ -40 & -44 & -20 \end{bmatrix}$$

so that

$$f(A) = \begin{bmatrix} -4 & -6 & -4 \\ 16 & 20 & 10 \\ -40 & -44 & -20 \end{bmatrix} + \begin{bmatrix} 0 & 0 & 4 \\ -16 & -24 & -16 \\ 64 & 80 & 40 \end{bmatrix} + \begin{bmatrix} 0 & 6 & 0 \\ 0 & 0 & 6 \\ -24 & -36 & -24 \end{bmatrix}$$

$$+ \begin{bmatrix} 4 & 0 & 0 \\ 0 & 4 & 0 \\ 0 & 0 & 4 \end{bmatrix} = \begin{bmatrix} 0 & 0 & 0 \\ 0 & 0 & 0 \\ 0 & 0 & 0 \end{bmatrix} .$$

Definition 1.11
If A is a matrix of order $n \times n$ and there exists another matrix B of order $n \times n$ such that

$$AB = BA = I$$

then B is called the **inverse** of A and is denoted by A^{-1}, and A is said to be **non-singular**. If A does not posess an inverse, we say that it is **singular**.

If the inverse of a matrix exists, it is unique.

Example 1.12
Find (if they exist) the inverses of

(i)
$$A = \begin{bmatrix} 1 & 2 \\ 3 & 4 \end{bmatrix} \text{ and}$$

(ii)
$$A = \begin{bmatrix} 1 & 2 \\ 3 & 6 \end{bmatrix} .$$

Solution

We must find $A^{-1} = \begin{bmatrix} u & v \\ w & z \end{bmatrix}$ such that

$$AA^{-1} = I = A^{-1}A .$$

(i)

$$\begin{bmatrix} 1 & 2 \\ 3 & 4 \end{bmatrix} \begin{bmatrix} u & v \\ w & z \end{bmatrix} = \begin{bmatrix} u & v \\ w & z \end{bmatrix} \begin{bmatrix} 1 & 2 \\ 3 & 4 \end{bmatrix} = \begin{bmatrix} 1 & 0 \\ 0 & 1 \end{bmatrix}.$$

On equating the elements (using the first product)

$$u + 2w = 1 \qquad v + 2z = 0$$

$$3u + 4w = 0 \qquad 3v + 4z = 1$$

so that $u = -2, v = 1, w = \frac{3}{2}$ and $z = -\frac{1}{2}$.

In this case A is non-singular.

(ii)

$$\begin{bmatrix} 1 & 2 \\ 3 & 6 \end{bmatrix} \begin{bmatrix} u & v \\ w & z \end{bmatrix} = \begin{bmatrix} 1 & 0 \\ 0 & 1 \end{bmatrix}$$

$$u + 2w = 1 \qquad v + 2z = 0$$

$$3u + 6w = 0 \qquad 3v + 6z = 1\,.$$

This system of equations is inconsistent, hence there is no solution. In this case A is a singular matrix.

Theorem 1.1

If $A = [a_{ij}]$ is a matrix of order $m \times n$, and $B = [b_{ij}]$ is a matrix of order $n \times r$, then

$$(AB)' = B'A'.$$

Proof

Let $AB = C = [c_{ij}]$

The $(i,j)^{\text{th}}$ element of $(AB)'$ is c_{ji}.

But $c_{ji} = \sum_{s=1}^{n} a_{js}\, b_{si} = \sum_{s=1}^{n} b'_{is}\, a'_{sj}$

where $A' = [a'_{ij}]$ and $B' = [b'_{ij}]$.

Also $B'A' = \sum_{s=1}^{n} b'_{is}a'_{sj}$, where $\sum_{s=1}^{n} b'_{is}a'_{sj}$ is the $(i,j)^{\text{th}}$ element of $B'A'$.

The result follows.

We can generalise the above result and show that

$$(A_1A_2 \ldots A_n)' = A'_n \ldots A'_2A'_1$$

provided that the products are defined.

Theorem 1.2

If the two matrices A and B, of order $n \times n$, are non-singular then

(i) AB is non-singular, and

(ii) $(AB)^{-1} = B^{-1}A^{-1}$.

Proof:

$$(AB)(B^{-1}A^{-1}) = A(BB^{-1})A^{-1} = (AI)A^{-1} = AA^{-1} = I.$$

Similarly $(B^{-1}A^{-1})(AB) = I$, hence (i) follows.

Also, since the inverse of a matrix is unique, (ii) follows. We can generalise the above result and show that if $A_1, A_2, \ldots A_n$ are non-singular matrices then

$$(A_1A_2 \ldots A_n)^{-1} = A_n^{-1} \ldots . A_2^{-1} A_1^{-1} .$$

Example 1.13

Given

$$A = \begin{bmatrix} 1 & -1 & 2 \\ 0 & 3 & -2 \end{bmatrix} \text{ and } B = \begin{bmatrix} 1 & 2 \\ -1 & 0 \\ 2 & 1 \end{bmatrix} .$$

find $(AB)'$ and $B'A'$.

Solution

Since

$$AB = \begin{bmatrix} 6 & 4 \\ -7 & -2 \end{bmatrix} \text{ then } (AB)' = \begin{bmatrix} 6 & -7 \\ 4 & -2 \end{bmatrix} .$$

Also

$$B'A' = \begin{bmatrix} 1 & -1 & 2 \\ 2 & 0 & 1 \end{bmatrix} \begin{bmatrix} 1 & 0 \\ -1 & 3 \\ 2 & -2 \end{bmatrix} = \begin{bmatrix} 6 & -7 \\ 4 & -2 \end{bmatrix}$$

Note that the inverse of a transposed non-singular square matrix A equals the transpose of its inverse. Indeed it is now simple to prove that

$$(A')^{-1} = (A^{-1})' .$$

1.3 PARTITIONED MATRICES

When dealing with matrices of high order, it is frequently useful to partition the matrices into a number of smaller arrays. Such partitioning can usually be done in a number of different ways. So although in this section no new theory is discussed, a useful technical device is introduced.

As an example, we consider the following matrix

$$A = \begin{bmatrix} 1 & 0 & 0 \\ 0 & 1 & 1 \\ -1 & 0 & 1 \\ 2 & 1 & 1 \end{bmatrix} .$$

One of the possible ways of partitioning A is the following:

$$A = \left[\begin{array}{cc:c} 1 & 0 & 0 \\ 0 & 1 & 1 \\ \hdashline -1 & 0 & 1 \\ 2 & 1 & 1 \end{array}\right] = \begin{bmatrix} A_{11} & A_{12} \\ A_{21} & A_{22} \end{bmatrix}$$

where

$$A_{11} = \begin{bmatrix} 1 & 0 \\ 0 & 1 \end{bmatrix} A_{21} = \begin{bmatrix} -1 & 0 \\ 2 & 1 \end{bmatrix} A_{12} = \begin{bmatrix} 0 \\ 1 \end{bmatrix} \text{ and } A_{22} = \begin{bmatrix} 1 \\ 1 \end{bmatrix}.$$

This particular partitioning has the merit of creating the unit submatrix A_{11} of order 2×2, which may be an advantage in a number of applications.

Example 1.14

Consider the matrix A defined above and the matrix

$$B = \begin{bmatrix} 1 & 0 & 0 & 0 \\ 0 & 0 & 1 & 0 \\ -1 & 0 & 0 & 1 \end{bmatrix}.$$

Evaluate the product AB.

Solution

We could use the partitioning of A defined above. We must now partition B in such a way that the partitioning of the rows of B corresponds to the partitioning of the columns of A. Thus B must be partitioned as follows:

$$B = \left[\begin{array}{cc:cc} 1 & 0 & 0 & 0 \\ 0 & 0 & 1 & 0 \\ \hdashline -1 & 0 & 0 & 1 \end{array}\right] = \begin{bmatrix} B_{11} & B_{12} \\ B_{21} & B_{22} \end{bmatrix}$$

where

$$B_{11} = \begin{bmatrix} 1 & 0 \\ 0 & 0 \end{bmatrix}, B_{12} = \begin{bmatrix} 0 & 0 \\ 1 & 0 \end{bmatrix}, B_{21} = [-1 \quad 0] \text{ and } B_{22} = [0 \quad 1].$$

We now have

$$AB = \begin{bmatrix} A_{11}B_{11} + A_{12}B_{21} & A_{11}B_{12} + A_{12}B_{22} \\ A_{21}B_{11} + A_{22}B_{21} & A_{21}B_{12} + A_{22}B_{22} \end{bmatrix}$$

$$= \begin{bmatrix} \begin{bmatrix} 1 & 0 \\ 0 & 1 \end{bmatrix}\begin{bmatrix} 1 & 0 \\ 0 & 0 \end{bmatrix} + \begin{bmatrix} 0 \\ 1 \end{bmatrix}[-1 \quad 0] & \begin{bmatrix} 1 & 0 \\ 0 & 1 \end{bmatrix}\begin{bmatrix} 0 & 0 \\ 1 & 0 \end{bmatrix} + \begin{bmatrix} 0 \\ 1 \end{bmatrix}[0 \quad 1] \\ \begin{bmatrix} -1 & 0 \\ 2 & 1 \end{bmatrix}\begin{bmatrix} 1 & 0 \\ 0 & 0 \end{bmatrix} + \begin{bmatrix} 1 \\ 1 \end{bmatrix}[-1 \quad 0] & \begin{bmatrix} -1 & 0 \\ 2 & 1 \end{bmatrix}\begin{bmatrix} 0 & 0 \\ 1 & 0 \end{bmatrix} + \begin{bmatrix} 1 \\ 1 \end{bmatrix}[0 \quad 1] \end{bmatrix}$$

$$= \begin{bmatrix} \begin{bmatrix} 1 & 0 \\ 0 & 0 \end{bmatrix} + \begin{bmatrix} 0 & 0 \\ -1 & 0 \end{bmatrix} & \begin{bmatrix} 0 & 0 \\ 1 & 0 \end{bmatrix} + \begin{bmatrix} 0 & 0 \\ 0 & 1 \end{bmatrix} \\ \begin{bmatrix} -1 & 0 \\ 2 & 0 \end{bmatrix} + \begin{bmatrix} -1 & 0 \\ -1 & 0 \end{bmatrix} & \begin{bmatrix} 0 & 0 \\ 1 & 0 \end{bmatrix} + \begin{bmatrix} 0 & 1 \\ 0 & 1 \end{bmatrix} \end{bmatrix}$$

$$= \begin{bmatrix} 1 & 0 & 0 & 0 \\ -1 & 0 & 1 & 1 \\ -2 & 0 & 0 & 1 \\ 1 & 0 & 1 & 1 \end{bmatrix}.$$

The above example may at first sight appear to demonstrate how simple it is to complicate life. In fact, when dealing with matrices of large order, the partitioning of matrices has distinct advantages. For example when using computers it can make use of much smaller storage space than in the case of direct matrix multiplication. There are also good reasons why partitioning is useful in a number of theoretical advances in matrix theory.

1.4 SOME SPECIAL MATRICES

Definition 1.12

(1) A square matrix A such that $A' = A$ is called **symmetric**.
(2) A square matrix A such that $A' = -A$ is called **skew-symmetric**.

For example

$$A_1 = \begin{bmatrix} 1 & 0 & -1 \\ 0 & 2 & 2 \\ -1 & 2 & 3 \end{bmatrix}, \qquad A_2 = \begin{bmatrix} 0 & -1 & 2 \\ 1 & 0 & -3 \\ -2 & 3 & 0 \end{bmatrix}.$$

A_1 is symmetric since each element $a_{ij} = a_{ji}$, and A_2 is skew-summetric, since each element $a_{ij} = -a_{ji}$. Note that the diagonal elements of a skew-symmetric matrix are zero.

Example 1.15
Show that if A is a square matrix
(i) $A + A'$ is symmetric, and
(ii) $A - A'$ is skew-symmetric.

Solution

(i) $(A+A')' = A' + (A')' = A' + A = A + A'$,

(ii) $(A-A')' = A' - (A')' = A' - A = -(A-A')$.

The results follow by Def. 1.12.

Although in the examples we have used up till now the matrices were over the real field (R) there is no reason why this should be so in general. We frequently have to work with matrices over the complex field (C).

Definition 1.13
When each element a_{ij} of a matrix A is replaced by its conjugate $\bar{a}_{ij}$, we obtain a matrix, denoted by $\bar{A}$, called **the conjugate** of A.

Definition 1.14
(1) A square matrix such that $(\bar{A})' = A$, is called **Hermitian**.
(2) If $A = -(\bar{A})'$, A is **skew-Hermitian**.

For example:

$$A_1 = \begin{bmatrix} 1 & 2-i \\ 2+i & -2 \end{bmatrix}, \qquad A_2 = \begin{bmatrix} 2i & 1-i \\ -1-i & 0 \end{bmatrix}.$$

Since the elements of A_1 are such that $a_{ij} = \bar{a}_{ji}$, A_1 is Hermitian. For A_2, $a_{ij} = -\bar{a}_{ji}$, hence it is skew-Hermitian. Note that the diagonal elements of a skew-Hermitian matrix must either be imaginary or zero.

Similar results to the ones established is Ex. 1.17 exist for Hermitian matrices. If A is a square matrix

(i) $A + A^*$ is Hermitian, and
(ii) $A - A^*$ is skew-Hermitian.

where A^* denotes the conjugate transpose of A, i.e. $A^* = (\bar{A})'$.

Definition 1.15

(1) A matrix A such that $A^2 = A$ is called **idempotent**.
(2) If A is a matrix and r is the least positive integer such that $A^{r+1} = A$, then A is called **periodic** of period r.
(3) If A is a matrix for which $A^r = 0$, A is called **nilpotent**. If r is the least positive integer for which this is true, A is said to be nilpotent of **order** r.

Example 1.16

Show that

$$A = \begin{bmatrix} 0 & 1 & 0 \\ 0 & 0 & 1 \\ -1 & -1 & -1 \end{bmatrix}$$

is periodic. Find the period.

Solution

We have

$$A^2 = \begin{bmatrix} 0 & 1 & 0 \\ 0 & 0 & 1 \\ -1 & -1 & -1 \end{bmatrix}\begin{bmatrix} 0 & 1 & 0 \\ 0 & 0 & 1 \\ -1 & -1 & -1 \end{bmatrix} = \begin{bmatrix} 0 & 0 & 1 \\ -1 & -1 & -1 \\ 1 & 0 & 0 \end{bmatrix}$$

$$A^3 = \begin{bmatrix} 0 & 0 & 1 \\ -1 & -1 & -1 \\ 1 & 0 & 0 \end{bmatrix}\begin{bmatrix} 0 & 1 & 0 \\ 0 & 0 & 1 \\ -1 & -1 & -1 \end{bmatrix} = \begin{bmatrix} -1 & -1 & -1 \\ 1 & 0 & 0 \\ 0 & 1 & 0 \end{bmatrix}$$

$$A^4 = \begin{bmatrix} -1 & -1 & -1 \\ 1 & 0 & 0 \\ 0 & 1 & 0 \end{bmatrix} \begin{bmatrix} 0 & 1 & 0 \\ 0 & 0 & 1 \\ -1 & -1 & -1 \end{bmatrix} = \begin{bmatrix} 1 & 0 & 0 \\ 0 & 1 & 0 \\ 0 & 0 & 1 \end{bmatrix}$$

therefore $A^5 = A$.

It follows that A is periodic having a period $r = 4$.

In control engineering theory, matrices are of a very special significance because we can represent the dynamic equation of a system in a vector-matrix notation, as explained in Sec. 1.5.

Definition 1.16

(1) The square matrix A is said to be **orthogonal** if

$$A'A = AA' = I.$$

(2) The square matrix A is said to be **unitary** if

$$A^*A = AA^* = I.$$

It is clear from these definitions that:
(a) both orthogonal and unitary matrices are necessarily non-singular,
(b) a real orthogonal matrix is unitary,
(c) the inverse of an orthogonal (unitary) matrix is equal to its transposed (transposed conjugate) matrix.

For example, consider

$$A_1 = \begin{bmatrix} \cos\theta & -\sin\theta \\ \sin\theta & \cos\theta \end{bmatrix}, \qquad A_2 = \begin{bmatrix} \frac{1}{3} & \frac{2}{3} & \frac{2}{3} \\ \frac{2}{3} & \frac{1}{3} & -\frac{2}{3} \\ -\frac{2}{3} & \frac{2}{3} & -\frac{1}{3} \end{bmatrix}$$

$$A_3 = \begin{bmatrix} \frac{1}{\sqrt{2}} & \frac{i}{\sqrt{2}} \\ \\ \frac{i}{\sqrt{2}} & \frac{1}{\sqrt{2}} \end{bmatrix}.$$

A_1 and A_2 are orthogonal and A_3 is unitary. Note that the sum of the squares of the elements of each row and column of an orthogonal matrix is unity. In Chapter 2, when operations on vectors are discussed in some detail, we would say that each vector is of **unit length**.

Also if $\mathbf{A}_i$ and $\mathbf{A}_j$ are any two columns of an orthogonal matrix A, then

$$\mathbf{A}_i'\mathbf{A}_j = 0 \qquad (i \neq j).$$

A similar property holds for the rows of A.

We would say (in Chapter 2) that any two row (or column) vectors of an orthogonal matrix are **mutually orthogonal**.

Vectors satisfying both the above properties, that is,

$$\mathbf{A}_i'\mathbf{A}_j = \delta_{ij} \qquad \text{(the Kronecker delta)}$$

are called **orthonormal**. If the elements of the vector $\mathbf{A}_i$ are complex numbers, the condition that it is of unit length is

$$\bar{\mathbf{A}}_i'\mathbf{A}_i = 1$$

where $\bar{\mathbf{A}}_i$ is the conjugate of $\mathbf{A}_i$, obtained by taking the conjugate of each complex element of $\mathbf{A}_i$. The condition for orthonormality becomes

$$\bar{\mathbf{A}}_i'\mathbf{A}_i = \delta_{ij} \quad .$$

Example 1.17

Prove that if A is skew-symmetric, then

$$(I+A)^{-1}(I-A)$$

is orthogonal (it can be proved that $I+A$ is non-singular)

Solution

Since A is skew-symmetric

$$(I-A)' = I+A,$$

hence

$$(I+A)^{-1}(I-A)[(I+A)^{-1}(I-A)]'$$
$$= (I+A)^{-1}(I-A)(I-A)'[(I+A)^{-1}]'$$
$$= (I+A)^{-1}(I+A)(I-A)(I-A)^{-1}$$

(since $(I-A)(I+A) = (I+A)(I-A)$ and $[(I+A)^{-1}]' = (I-A)^{-1}$).

$$= I.$$

The result follows.

1.5 THE STATE-SPACE CONCEPT

When considering the dynamic behaviour of a system it is frequently convenient to classify the various variables involved into three categories; input, output and state. Schematically we could represent the situation as follows

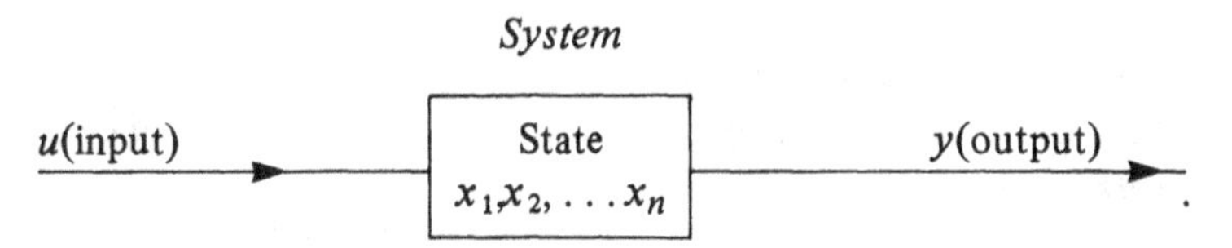

A more general situation would make allowance for multi-input and multi-output variables.

The output $y(t)$, of an nth-order linear system, having an input $u(t)$, can be determined for $t \geq 0$ from a differential equation of the form

$$\overset{(n)}{y} + a_1\overset{(n-1)}{y} + \ldots + a_{n-1}\dot{y} + a_n y = u \tag{1.1}$$

where $\overset{(m)}{y} = \dfrac{d^m y}{dt^m}$

if $u(t)$ and $y(0), \dot{y}(0), \ldots \overset{(n-1)}{y}(0)$ are known.

If we define $x_1 = y, x_2 = \dot{y}, \ldots x_n = \overset{(n-1)}{y}$, then we can write Eq. (1.1) as n simultaneous differential equations, each of order 1:

$$\dot{x}_1 = x_2$$

$$\dot{x}_2 = x_3$$

$$\vdots$$

$$\dot{x}_{n-1} = x_n$$

and from (1.1) $\dot{x}_n = -a_n x_1 - a_{n-1}x_2 - \ldots - a_1 x_n + u$
which can be written in matrix form as

$$\begin{bmatrix} \dot{x}_1 \\ \dot{x}_2 \\ \vdots \\ \dot{x}_{n-1} \\ \dot{x} \end{bmatrix} = \begin{bmatrix} 0 & 1 & \dots & 0 & 0 \\ 0 & 0 & \dots & 0 & 0 \\ \vdots & & & & \\ 0 & 0 & \dots & 0 & 1 \\ -a_n & -a_{n-1} & & -a_2 & -a_1 \end{bmatrix} \begin{bmatrix} x_1 \\ x_2 \\ \vdots \\ x_{n-1} \\ x_n \end{bmatrix} + \begin{bmatrix} 0 \\ 0 \\ \vdots \\ 0 \\ 1 \end{bmatrix} [u] \quad (1.2)$$

that is, as

$$\dot{\mathbf{x}} = A\mathbf{x} + Bu$$

where $\mathbf{x}$, A, and B are defined in (1.2).

The output of the system is y (which we defined as x_1) and can be written in matrix form as

$$y = [1 \quad 0 \quad 0 \quad \dots \quad 0] \begin{bmatrix} x_1 \\ x_2 \\ \vdots \\ x_n \end{bmatrix}, \quad (1.3)$$

that is,

$$y = C\mathbf{x}$$

where C is defined in (1.3).

The vector $\mathbf{x}$ is called the **state vector** of the system.
Eqs. (1.2) and (1.3) in the form

$$\begin{aligned} \dot{\mathbf{x}} &= A\mathbf{x} + Bu \\ y &= C\mathbf{x} \end{aligned} \quad (1.4)$$

are known as the **state equations** of the system considered.

The collection of all possible values which $\mathbf{x}$ can assume at time t, forms the **state-space** of the system.

Example 1.18
Obtain the state equations of the system defined by

$$\dddot{y} - 2\ddot{y} + \dot{y} - 2y = u \tag{1.5}$$

Solution
Let $x_1 = y$, $x_2 = \dot{y}$, and $x_3 = \ddot{y}$, then

$$\dot{x}_1 = x_2$$
$$\dot{x}_2 = x_3$$
$$\dot{x}_3 = 2x_1 - x_2 + 2x_3 + u,$$

that is,

$$\begin{bmatrix} \dot{x}_1 \\ \dot{x}_2 \\ \dot{x}_3 \end{bmatrix} = \begin{bmatrix} 0 & 1 & 0 \\ 0 & 0 & 1 \\ 2 & -1 & 2 \end{bmatrix} \begin{bmatrix} x_1 \\ x_2 \\ x_3 \end{bmatrix} + \begin{bmatrix} 0 \\ 0 \\ 1 \end{bmatrix} [u]$$

and

$$y = [1 \quad 0 \quad 0] \begin{bmatrix} x_1 \\ x_2 \\ x_3 \end{bmatrix}.$$

The choice of state variables is not unique, indeed there exist a limitless number of possibilities. The choice made above leads to a convenient form of the matrix A in the state equations, but this form is not necessarily the best in all circumstances. For example $\dot{x}_3$ is seen to be a function of all three state variables. In many applications it is very useful to have the state variables **decoupled**, which is achieved when each $\dot{x}_i$ is a function of x_i (and the input) only. Such a decoupling implies that the matrix A in the state equations is diagonal.

Example 1.19
Choose appropriate state variables for the system defined by (1.5) in Ex. 1.18 so that the corresponding matrix A in the state equations is in a diagonal form.

Solution
If we choose

$$z_1 = \frac{1}{5}y + \frac{1}{5}\ddot{y}$$

$$z_2 = \frac{1}{5}(2+i)y - \frac{1}{2}i\dot{y} + \frac{1}{10}(-1+2i)\ddot{y} \tag{1.6}$$

$$z_3 = \frac{1}{5}(2-i)y + \frac{1}{2}i\dot{y} - \frac{1}{10}(1+2i)\ddot{y}.$$

then

$$\dot{z}_1 = \frac{2}{5}(\ddot{y} + y) - \frac{1}{5}u \text{ (using the fact that } \dddot{y} = 2\ddot{y} - \dot{y} + 2y - u)$$

$$= 2z_1 + \frac{1}{5}u.$$

Similarly

$$\dot{z}_2 = iz_2 + \frac{1}{10}(-1+2i)u$$

and

$$\dot{z}_3 = -iz_3 - \frac{1}{10}(1+2i)u.$$

In matrix form this is written as

$$\begin{bmatrix} \dot{z}_1 \\ \dot{z}_2 \\ \dot{z}_3 \end{bmatrix} = \begin{bmatrix} 2 & 0 & 0 \\ 0 & i & 0 \\ 0 & 0 & -i \end{bmatrix} \begin{bmatrix} z_1 \\ z_2 \\ z_3 \end{bmatrix} + \frac{1}{10} \begin{bmatrix} +2 \\ -1+2i \\ -1-2i \end{bmatrix} [u]$$

and (on addition of Eqs (1.6))

$$y = z_1 + z_2 + z_3,$$

that is,

$$y = [1 \quad 1 \quad 1] \begin{bmatrix} z_1 \\ z_2 \\ z_3 \end{bmatrix}.$$

Examples 1.18 and 1.19 show how the form of the matrix A of the state equations depends on the choice of the state variables. The rational for the choice of Eqs (1.6) will become clear when we have obtained the state equations as in Ex. 1.18, we can apply any (non-singular) transformation $\mathbf{x} = P\mathbf{z}$. The state equations then become

$$P\dot{\mathbf{z}} = AP\mathbf{z} + Bu$$

so that

$$\dot{\mathbf{z}} = P^{-1}AP\mathbf{z} + P^{-1}Bu$$

and

$$y = CP\mathbf{z}.$$

We have chosen P such that $P^{-1}AP$ is diagonal. The form of the matrix A in Ex. 1.18 (called **companion** form), leads us to choose P to be the Vandermonde matrix of the eigenvalues of A as explained in Chapter 8 (Sec. 8.2).

Problems for Chapter 1

1) For each pair of matrices find, when the operation is defined (i) $A+B$, (ii) $A-2B$, (iii) AB and (iv) BA

(a) $$A = \begin{bmatrix} 2 & -1 \\ -1 & 3 \end{bmatrix}; \qquad B = \begin{bmatrix} 1 & 2 \\ 2 & 0 \end{bmatrix}.$$

(b) $$A = \begin{bmatrix} 1 & 2 & -2 \\ -1 & 0 & 1 \end{bmatrix}; \qquad B = \begin{bmatrix} 2 & -2 \\ 0 & 2 \\ 1 & 1 \end{bmatrix}.$$

(c) $$A = \begin{bmatrix} 1 & 0 & -1 \\ 0 & 1 & 0 \\ 2 & 0 & 0 \end{bmatrix}, \qquad B = \begin{bmatrix} 1 & 2 & 0 \\ -2 & 1 & 0 \\ 0 & 0 & 2 \end{bmatrix}.$$

2) Let

$$A = \begin{bmatrix} 1 & 1 & 1 & 1 \\ 0 & -1 & -2 & -3 \\ 0 & 0 & 1 & 3 \\ 0 & 0 & 0 & -1 \end{bmatrix}.$$

Show that A is idempotent.

3) A is an idempotent matrix of order 2×2 and Λ_1, Λ_2 are 2×2 diagonal matrices.

If $P = A\Lambda_1 A$ and $Q = A\Lambda_2 A$, prove that

$$PQ = QP.$$

4) (a) Show that any matrix A can be written as the sum of a symmetric matrix and a skew-symmetric matrix.

(b) Write

$$A = \begin{bmatrix} 1 & \frac{1}{2} & -2 \\ \frac{5}{2} & 0 & \frac{1}{4} \\ 2 & 2 & -1 \end{bmatrix}$$

as the sum of a symmetric and of a skew-symmetric matrices.

5) Classify the following matrices as symmetric, skew-symmetric, Hermitian or skew-Hermitian.

(i) $\begin{bmatrix} 5 & -4-2i \\ -4+2i & 11 \end{bmatrix}$;

(ii) $\begin{bmatrix} 0 & 2i \\ 2i & 0 \end{bmatrix}$;

(iii) $\begin{bmatrix} 2 & -1 \\ -1 & 3 \end{bmatrix}$;

(iv) $\begin{bmatrix} 2i & 2i \\ 2i & 0 \end{bmatrix}$;

(v) $\begin{bmatrix} 0 & -2 \\ 2 & 0 \end{bmatrix}$;

(vi) $\begin{bmatrix} 1 & 2 \\ 2 & 3 \end{bmatrix}$;

6) Using the partitioning shown, evaluate the product AB,

$$A = \left[\begin{array}{ccc|c} 1 & 0 & 0 & 1 \\ 0 & 1 & 0 & 1 \\ 0 & 0 & 1 & 1 \\ \hline 0 & 0 & 0 & 1 \end{array}\right] \quad \text{and } B = \left[\begin{array}{ccc|c} 1 & 0 & 0 & 0 \\ 0 & 2 & 0 & 0 \\ 0 & 0 & 3 & 1 \\ \hline 1 & 0 & 0 & 0 \end{array}\right].$$

7) Show that the matrix

$$A = \begin{bmatrix} 1 & 5 & -2 \\ 1 & 6 & -1 \\ 3 & 2 & -3 \end{bmatrix}$$

is nilpotent and find its order.

8) Given $A = \begin{bmatrix} a & b \\ c & d \end{bmatrix}$ and $ad-bc \neq 0$

show that

$$A^{-1} = \frac{1}{ad-bc}\begin{bmatrix} d & -b \\ -c & a \end{bmatrix}.$$

9) The voltage-current relationship for a series impedance Z

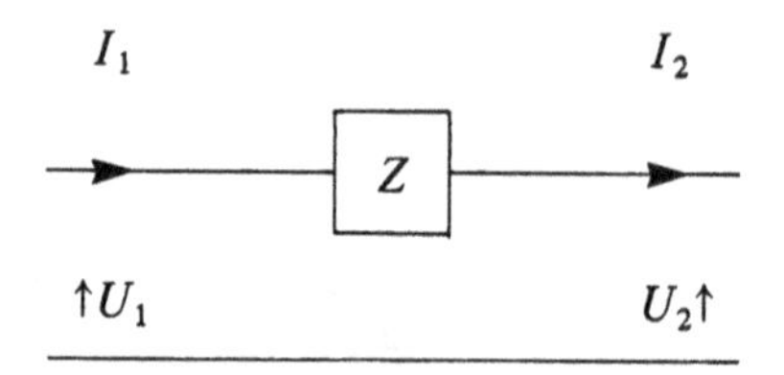

is $\begin{bmatrix} U_1 \\ I_1 \end{bmatrix} = \begin{bmatrix} 1 & Z \\ 0 & 1 \end{bmatrix}\begin{bmatrix} U_2 \\ I_2 \end{bmatrix}$

and for a shunt impedance

I_1 I_2

$U_1\uparrow$ Z $U_2\uparrow$

it is $\begin{bmatrix} U_1 \\ I_1 \end{bmatrix} = \begin{bmatrix} 1 & 0 \\ 1/Z & 1 \end{bmatrix}\begin{bmatrix} U_2 \\ I_2 \end{bmatrix}.$

Determine the voltage–current relationships for the two networks below:

(a)

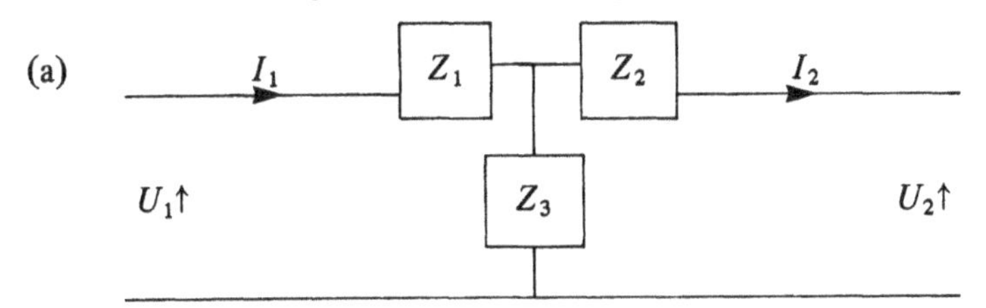

(b)

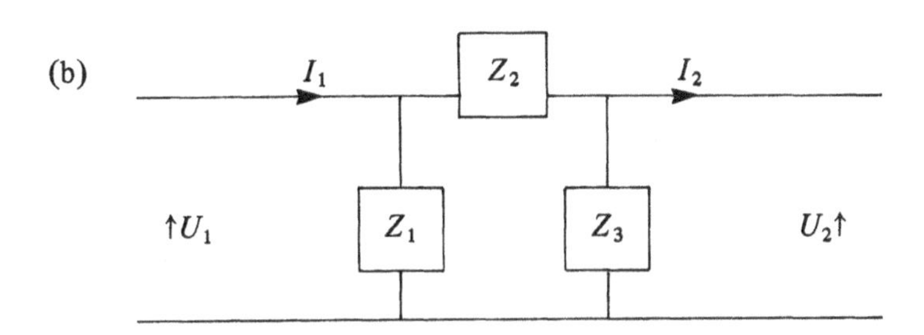

10) The dynamic behaviour of a system with input u and output y is determined by

$$\dddot{y} + 6\ddot{y} + 11\dot{y} + 6y = u.$$

Write this equation in a state-space form.

11) To describe electron spin, the following three matrices, called Pauli matrices, are used

$$X_1 = \frac{1}{2}\begin{bmatrix} 0 & 1 \\ 1 & 0 \end{bmatrix}, \quad X_2 = \frac{1}{2}\begin{bmatrix} 0 & -i \\ i & 0 \end{bmatrix} \text{ and } X_3 = \frac{1}{2}\begin{bmatrix} 1 & 0 \\ 0 & -1 \end{bmatrix},$$

(where the Planck's constant is taken as unity).
Find

(i) $X_i^2 \quad (i = 1, 2, 3)$

(ii) $X_iX_j + X_jX_i \quad (i, j = 1, 2, 3)$.

12) Find the condition that must be satisfied by the real numbers p and q so that the matrix

$$A = \begin{bmatrix} p-q & q+p \\ p+q & q-p \end{bmatrix}$$

is orthogonal.

CHAPTER 2

Vector Spaces

2.1 VECTORS

It is generally accepted in elementary physics that a vector is an entity specified by a line segment having a magnitude and a direction. If we consider a vector **a** in a plane we could represent it by the directed line segment OP with respect to a coordinate system having axes Ox and Oy. If the point P has coordinates (x,y) we could specify the vector **a** by the ordered pair (x,y). Similarly we could specify a vector in three dimensions by the ordered triad (x,y,z) where the coordinate of the point P is (x,y,z).

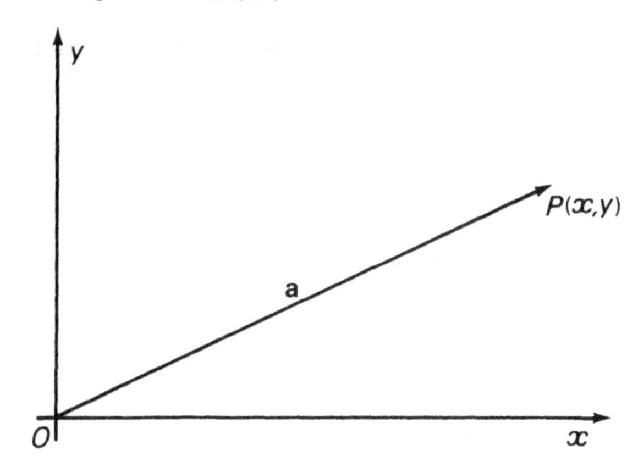

An extension of this concept to n-dimensions leads to the following definition.

Definition 2.1

A vector **x** of order n is an ordered n-tuple of numbers $(x_1 x_2, \ldots x_n)$.

To be precise we would need to specify that the n numbers $x_1, x_2, \ldots x_n$ belong to a particular field F. We would then say that the vector **x** is over the field F. Generally we consider vectors over the real or complex fields.

The numbers $x_1, x_2, \ldots x_n$ are called the **components** of the vector.

Definition 2.2
Two vectors $\mathbf{x} = (x_1,x_2,\ldots x_n)$ and $\mathbf{y} = (y_1,y_2,\ldots y_n)$ are **equal** if and only if their corresponding components are equal, that is, if and only if

$$x_i = y_i \quad (i = 1,2,\ldots n).$$

Definition 2.3
The **zero** vector **O** is the vector having all its components equal to zero, that is

$$\mathbf{O} = (0,0,\ldots 0).$$

Definition 2.4
(i) **Addition** of two vectors **x** and **y** of the same order is defined by

$$\mathbf{x} + \mathbf{y} = (x_1,x_2,\ldots x_n) + (y_1,y_2,\ldots y_n) = (x_1 + y_1,x_2 + y_2,\ldots x_n + y_n)$$

(ii) **Multiplication** of a vector **x** by a scalar k is defined by

$$k\mathbf{x} = k(x_1,x_2,\ldots x_n) = (kx_1,kx_2,\ldots kx_n).$$

We refer to the multiplication of a vector by a scalar as **scalar multiplication**.

(iii) There are two types of multiplication of one vector by another, both of the same order.

(a) The **inner product** (also called the **dot product** or the **scalar product** – not to be confused with the multiplication of a vector by a scalar!).

If $\mathbf{x} = (x_1,x_2,\ldots x_n)$ and $\mathbf{y} = (y_1,y_2,\ldots y_n)$ then the inner product is defined as

$$\mathbf{x} \cdot \mathbf{y} = x_1y_1 + x_2y_2 + \ldots + x_ny_n .$$

(the same rule as the multiplication of a row matrix **x** by a column matrix **y**).

For example if $\mathbf{x} = (1,2,3)$ and $\mathbf{y} = (-1,1,2)$

$$\mathbf{x} \cdot \mathbf{y} = 1 \cdot (-1) + 2 \cdot 1 + 3 \cdot 2 = 7.$$

Notice that the inner product of two vectors is a scalar.

(b) The **vector product** (also known as the **cross product**) of two vectors $\mathbf{x} = (x_1,x_2,x_3)$ and $\mathbf{y} = (y_1,y_2,y_3)$ is defined as

$$\mathbf{x} \times \mathbf{y} = (x_2y_3 - y_2x_3, x_3y_1 - y_3x_1, x_1y_2 - y_1x_2).$$

For the above example

$$\mathbf{x} \times \mathbf{y} = (2.2 - 1.3, 3 . (-1) - 2.1, 1.1 - (-1).2) = (1,-5,3)$$

Notice that the vector product of two vectors is itself a vector.

From the above definitions it is not difficult to prove the following:

If $\mathbf{x}$, $\mathbf{y}$ and $\mathbf{z}$ are vectors of the same order and α, β are scalars;

(1) $\mathbf{x} + \mathbf{y} = \mathbf{y} + \mathbf{x}$
(2) $\mathbf{x} + (\mathbf{y}+\mathbf{z}) = (\mathbf{x}+\mathbf{y}) + \mathbf{z}$
(3) $\alpha(\mathbf{x}+\mathbf{y}) = \alpha\mathbf{x} + \alpha\mathbf{y}$
(4) $(\alpha+\beta)\mathbf{x} = \alpha\mathbf{x} + \beta\mathbf{x}$
(5) $(\alpha\beta)\mathbf{x} = \alpha(\beta\mathbf{x})$
(6) $(-1)\mathbf{x} = (-x_1, -x_2, \ldots -x_n) = -\mathbf{x}$

All the above definitions have simple geometric interpretations. For example the following represents the addition of the two vectors **x** and **y**.

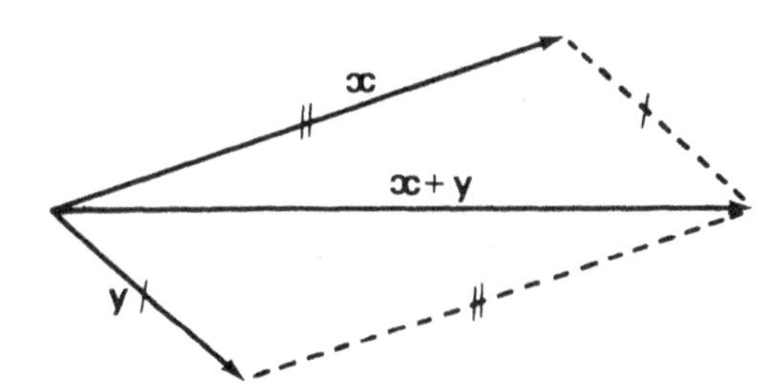

Because the definition of a 'vector' is so general it applies to entities arising in various problems, even if at first sight it is not possible to associate such entities with directed segments (or 'arrows'). Having defined various desirable operations on vectors we are in a position to summarise them in a definition of a structure known as a **vector space**.

Definition 2.5
A **vector space** V over a field F is a set of elements (called **vectors**) together with two operations, addition and scalar multiplication, satisfying the following conditions:

Let $\mathbf{x}, \mathbf{y}, \mathbf{z} \in V$ and $\alpha, \beta \in F$

(1) $(\mathbf{x}+\mathbf{y})$ is a unique vector $\in V$ (we call this property 'closure' with respect to addition).
(2) $\mathbf{x} + \mathbf{y} = \mathbf{y} + \mathbf{x}$.

(3) $(\mathbf{x}+\mathbf{y}) + \mathbf{z} = \mathbf{x} + (\mathbf{y}+\mathbf{z})$.
(4) $\exists$ a unique element $\mathbf{O} \in V$ such that $\mathbf{x} + \mathbf{O} = \mathbf{O} + \mathbf{x} = \mathbf{x}$, all $\mathbf{x} \in V$.
(5) For each $\mathbf{x} \in V$, $\exists(-\mathbf{x}) \in V$ such that $\mathbf{x} + (-\mathbf{x}) = \mathbf{O}$.
(6) The scalar product $\alpha\mathbf{x}$ is a unique element in V.
(7) $\alpha(\beta\mathbf{x}) = (\alpha\beta)\mathbf{x}$.
(8) $\alpha(\mathbf{x}+\mathbf{y}) = \alpha\mathbf{x} + \alpha\mathbf{y}$.
(9) $(\alpha+\beta)\mathbf{x} = \alpha\mathbf{x} + \beta\mathbf{x}$.
(10) $1.\mathbf{x} = \mathbf{x}$ (1 is the unit element in F).

Any entities satisfying the above axioms are called **vectors**.

Example 2.1
Show that the set P_n of all polynomials of degree n or less is a vector space.

Solution
A polynomial $P(t)$ of degree n has the form

$$P(t) = a_0t^n + a_1t^{n-1} + \ldots + a_n.$$

Such a polynomial is fully defined by the $(n+1)$-tuple of numbers (the polynomial coefficients)

$$(a_0, a_1, \ldots a_n).$$

Thus the $(n+1)$-tuple

$$(b_0, b_1, \ldots b_n)$$

uniquely defines the polynomial

$$q(t) = b_0t^n + b_1t^{n-1} + \ldots + b_n.$$

The sum of the two polynomials is uniquely defined by the $(n+1)$-tuple

$$(a_0+b_0, a_1+b_1, \ldots a_n+b_n).$$

It is now simple to show that all the axioms for a vector space are satisified.

Note that the sum of the $(n+1)$-tuples

$$(a_0, a_1, \ldots a_n) + (-1)(a_0, b_1, \ldots b_n)$$

$$= (0, a_1-b_1, \ldots a_n-b_n)$$

and defines a polynomial of degree $(n-1)$ (if $a_1 \neq b_1$). This is why it is the set of polynomials of degree n and *less* that is a vector space, whereas the set of polynomials of degree n is *not* a vector space.

Of course it goes without saying that n-tuples of numbers, that is vectors together with the two operations of vector addition and scalar multiplication, form a vector space. If the elements of the *n-tuples are real numbers,* we denote this *vector space by* R^n *and call it* the **n-dimensional space**.

Another example of a vector space is the set of all matrices of order $m \times n$ together with the operations of matrix addition and scalar multiplication. We have defined (Def. 2.1) a vector to be an n-tuple of elements from a field. Perhaps it would be more precise to have called n-tuples of elements 'coordinate vectors' or some other name. This is because elements of a vector space are called 'vectors'. For example all matrices of order $m \times n$ are 'vectors'. But in spite of possible confusion it is generally clear from the context what 'vectors' are being considered, and in general there is little need to be more specific.

Definition 2.6

A non-empty subset W of a vector space V is a **subspace** of V if

(i) W is closed under vector addition; that is, if $\mathbf{x}, \mathbf{y} \in W$, then $\mathbf{x} + \mathbf{y} \in W$,

and

(ii) W is closed under scalar multiplicity; that is, if $\mathbf{x} \in W$ and $\alpha \in F$ then $\alpha\mathbf{x} \in W$.

Conditions (i) and (ii) above can be combined in a single condition:

If $\mathbf{x}, \mathbf{y} \in W$ and $\alpha, \beta \in F$, then

$$\alpha\mathbf{x} + \beta\mathbf{y} \in W.$$

As a direct consequence of this definition, it is simple to prove that W together with the two operation defined for V is a vector space in its own right.

Definition 2.7

The expression $\alpha\mathbf{x} + \beta\mathbf{y}$ is called a **linear combination** of $\mathbf{x}$ and $\mathbf{y}$. More generally we define the linear combination of the n vectors $\mathbf{x}_1, \mathbf{x}_2, \ldots \mathbf{x}_n$ is an expression of the form

$$\alpha\mathbf{x}_1 + \beta\mathbf{x}_2 + \ldots + \omega\mathbf{x}_n$$

where $\alpha, \beta, \ldots \omega \in F$.

Example 2.2
Let V be the vector space R^3. Determine whether W is a subspace of V if W consists of the following vectors:

(i) $W = \{(a,b,c): b < 0\}$.
(ii) $W = \{(a,b,c): a + b + c = 1\}$.
(iii) $W = \{(a,b,c): b = 2c\}$.

Solution

(i) W consists of all vectors in R^3 such that the second component is negative. So, for example, $\mathbf{x} = (1,-1,2) \in W$.
Let $\alpha = -2 \in R$, then $\alpha\mathbf{x} = (-2,2,-4)$ which $\notin W$ since the second component 2 is positive. Hence W is not a subspace of V.

(ii) Assume that $\mathbf{x} = (a,b,c) \in W$, then $a + b + c = 1$.
Let $\alpha = 2 \in R$, then $\alpha\mathbf{x} = (\alpha a, \alpha b, \alpha c)$
and $\alpha a + \alpha b + \alpha c = \alpha(a+b+c) = 2$. Hence $\alpha\mathbf{x} \notin W$.
It follows that W is not a subspace of V.

(iii) We shall see in a moment that if W is a subspace of V, the $\mathbf{O}$ (the null vector) $\in W$. In this example

$$\mathbf{O} = (0,0,0) \in W \text{ as } b = 2c \text{ is satisfied since } b = c = 0.$$

If $\mathbf{x}_1 = (a_1,b_1,c_1)$ and $\mathbf{x}_2 = (a_2,b_2,c_2)$ both $\in W$, then $b_1 = 2c_1$ and $b_2 = 2c_2$. Let α and β be any scalars, then

$$\begin{aligned}\alpha\mathbf{x}_1 + \beta\mathbf{x}_2 &= \alpha(a_1,b_1,c_1) + \beta(a_2,b_2,c_2)\\ &= (\alpha a_1,\alpha b_1,\alpha c_1) + (\beta a_2+\beta b_2+\beta c_2)\\ &= (\alpha a_1+\beta a_2,\alpha b_1+\beta b_2,\alpha c_1+\beta c_2).\end{aligned}$$

But $\alpha b_1 + \beta b_2 = \alpha(2c_1) + \beta(2c_2) = 2(\alpha c_1+\beta c_2)$, hence $\alpha\mathbf{x}_1 + \beta\mathbf{x}_2 \in W$. It follows that W is a subspace of V.

Example 2.3
Write the vector $\mathbf{z} = (-2,1,3)$ as a linear combination of the following vectors:

(i) $\mathbf{x}_1 = (1,1,1)$ $\mathbf{x}_2 = (1,1,0)$ and $\mathbf{x}_3 = (1,0,0)$, and
(ii) $\mathbf{y}_1 = (2,1,0)$ $\mathbf{y}_2 = (-1,2,1)$ and $\mathbf{y}_3 = (1,0,1)$.

Solution

(i) We must find the unknown scalars α, β and γ so that

$$\mathbf{z} = \alpha\mathbf{x}_1 + \beta\mathbf{x}_2 + \gamma\mathbf{x}_3.$$

Thus $(-2,1,3) = (\alpha,\alpha,\alpha) + (\beta,\beta,0) + (\gamma,0,0)$

$$= (\alpha+\beta+\gamma,\alpha+\beta,\alpha).$$

Equating the components of the two vectors, we have

$$-2 = \alpha + \beta + \gamma$$

$$1 = \alpha + \beta$$

$$3 = \alpha.$$

On solving this system of equations, we find

$$\alpha = 3, \beta = -2, \gamma = -3.$$

Hence $\mathbf{z} = 3\mathbf{x}_1 - 2\mathbf{x}_2 - 3\mathbf{x}_3$

(ii) This time find α, β and γ so that

$$\mathbf{z} = \alpha\mathbf{y}_1 + \beta\mathbf{y}_2 + \gamma\mathbf{y}_3, \text{ that is,}$$

$$(-2,1,3) = (2\alpha,\alpha,0) + (-\beta,2\beta,\beta) + (\gamma,0,\gamma)$$

$$= (2\alpha-\beta+\gamma,\alpha+2\beta,\beta+\gamma),$$

so that

$$-2 = 2\alpha - \beta + \gamma$$

$$1 = \alpha + 2\beta$$

$$3 = \beta + \gamma.$$

On solving, we find $\alpha = -\frac{4}{3}, \beta = \frac{7}{6}$ and $\gamma = \frac{11}{6}$,

hence

$$\mathbf{z} = -\frac{4}{3}\mathbf{y}_1 + \frac{7}{6}\mathbf{y}_2 + \frac{11}{6}\mathbf{y}_3.$$

It was mentioned in the solution to Ex. 2.2 (iii) that if W is a subspace of V then $\mathbf{O} \in W$. This is true because if W is a subspace of V, by definition it must be a vector space and so must satisfy all the axioms of Def. 2.5, in particular axiom 4. Of course if we take $\alpha = 0$ in Def. 2.6 we also find that $0.\mathbf{x} = \mathbf{O} \in W$. Generally it is worthwhile to begin an investigation whether W is a subspace of V by verifying whether or not $\mathbf{O} \in W$.

In Ex. 2.2 (i) we could have stated immediately that W is not a subspace of V since $\mathbf{O} \notin W$.

It is worthwhile noting that the set $\{\mathbf{O}\}$ involving only the one vector $\mathbf{O}$ is a vector space in its own right. In fact it is the only vector space having a finite number of elements. On the other hand we shall be concerned with vector spaces which are completely described by a finite number of vectors which we say, form a **basis** for the vector space.

Theorem 2.1
Let S be a subset of a vector space V, the set of all linear combinations of vectors in S, denoted by $L(S)$, is a subspace of V.

Proof
Let $\mathbf{x}, \mathbf{y} \in L(S)$, then

$$\mathbf{x} = \alpha_1 \mathbf{x}_1 + \alpha_2 \mathbf{x}_2 + \ldots + \alpha_n \mathbf{x}_n, \text{ and}$$

$$\mathbf{y} = \beta_1 \mathbf{y}_1 + \beta_2 \mathbf{y}_2 + \ldots + \beta_m \mathbf{y}_m$$

where $\mathbf{x}_i, \mathbf{y}_j \in S$ and α_i, β_j are scalars.

Then for any scalars a, b

$$a\mathbf{x} + b\mathbf{y} = a\,\alpha_1\mathbf{x}_1 + \ldots + a\alpha_n\mathbf{x}_n + b\beta_1\mathbf{y}_1 + \ldots + b\beta_m\mathbf{y}_m.$$

Hence $a\mathbf{x} + b\mathbf{y} \in L(S)$ (since it is a linear combination of $\mathbf{x}_i$ and $\mathbf{y}_j \in S$). It follows that $L(S)$ is a subspace of V.

Definition 2.8
The subspace $L(S)$ defined above is said to be **spanned** (or **generated**) by S. The vectors in S are called the **generators** of $L(S)$.

Example 2.4
Show that
(i) The vectors $\mathbf{x}_1 = (1,1,1)$, $\mathbf{x}_2 = (1,1,0)$ and $\mathbf{x}_3 = (1,0,0)$ span R^3.
(ii) The polynomials $f_1(t) = t^2 + 2t - 1$, $f_2(t) = t + 1$, $f_3(t) = 1$ span the space of polynomials of degree ≤ 2.

Solution

(i) We must show that an arbitrary vector $(a,b,c) \in R^3$ is a linear combination of $\mathbf{x}_1$, $\mathbf{x}_2$ and $\mathbf{x}_3$, that is, that

$$(a,b,c) = \alpha\mathbf{x}_1 + \beta\mathbf{x}_2 + \gamma\mathbf{x}_3 = (\alpha,\alpha,\alpha) + (\beta,\beta,0) + (\gamma,0,0)$$

so that $a = \alpha + \beta + \gamma$

$b = \alpha + \beta$ that is, $\alpha = c$, $\beta = b-c$, and $\gamma = a-b$

$c = \alpha$

hence

$$(a,b,c) = c\mathbf{x}_1 + (b-c)\mathbf{x}_2 + (a-b)\mathbf{x}_3.$$

(ii) Let $P_2(t) = at^2 + bt + c$ be any polynomial of degree 2 or less (that is, a or b or c could be 0).
If $at^2 + bt + c = \alpha f_1(t) + \beta f_2(t) + \gamma f_3(t)$, then $(a,b,c) = \alpha(1,2,-1) + \beta(0,1,1) + \gamma(0,0,1)$ so that $a = \alpha$, $b = 2\alpha + \beta$ and $c = -\alpha + \beta + \gamma$, that is, $\alpha = a$, $\beta = b-2a$, and $\gamma = 3a - b + c$, hence $P_2 = af_1 + (b-2a)f_2 + (3a-b+c)f_3$.

2.2 LINEAR DEPENDENCE AND BASES

Definition 2.9

Let $\mathbf{x}_1, \mathbf{x}_2, \ldots \mathbf{x}_r$ be vectors $\in V$ (that is, elements of a vector space V) and $\alpha_1, \alpha_2, \ldots \alpha_r$ be scalars. We consider the equation

$$\alpha_1\mathbf{x}_1 + \alpha_2\mathbf{x}_2 + \ldots + \alpha_r\,\mathbf{x}_r = \mathbf{0}.$$

If at least one $\alpha_i \neq 0$ in the above equation, the vectors $\mathbf{x}_1, \mathbf{x}_2, \ldots \mathbf{x}_r$ are said to be **linearly dependent** and the equation is **non-trivial.** If, on the other hand, the above equation is satisfied only when $\alpha_1 = \alpha_2 = \ldots = \alpha_r = 0$, the vectors are said to be **linearly independent**, and this solution is said to be **trivial.**

Example 2.5

Determine whether the following sets of vectors are linearly dependent or independent

(i) $\mathbf{x}_1 = (2,1,3), \mathbf{x}_2 = (1,0,2), \mathbf{x}_3 = (1,2,0).$

(ii) $\mathbf{x}_1 = (1,2,3), \mathbf{x}_2 = (1,1,0), \mathbf{x}_3 = (1,0,1).$

Solution

(i) We must determine whether at least one $\alpha_i \neq 0$ in the expression

$$\alpha_1\mathbf{x}_1 + \alpha_2\mathbf{x}_2 + \alpha_3\mathbf{x}_3 = \mathbf{0}, \text{ that is,}$$

$$(2\alpha_1,\alpha_1,3\alpha_1) + (\alpha_2,0,2\alpha_2) + (\alpha_3,2\alpha_3,0) = (0,0,0)$$

so that

$$2\alpha_1 + \alpha_2 + \alpha_3 = 0$$

$$\alpha_1 + 0 + 2\alpha_3 = 0$$

$$3\alpha_1 + 2\alpha_2 + 0 = 0.$$

On solving this set of homogeneous equations, we obtain $(\alpha_1,\alpha_2,\alpha_3) = (2r,-3r,-r)$, r being any real number. So in this case

$$2\mathbf{x}_1 - 3\mathbf{x}_3 - \mathbf{x}_3 = \mathbf{0}$$

and $\mathbf{x}_1, \mathbf{x}_2, \mathbf{x}_3$ are linearly dependent.

(ii) $\alpha_1\mathbf{x}_1 + \alpha_2\mathbf{x}_2 + \alpha_3\mathbf{x}_3 = (\alpha_1,2\alpha_1,3\alpha_1) + (\alpha_2,\alpha_2,0) + (\alpha_3,0,\alpha_3) = (0,0,0).$

Hence,

$$\alpha_1 + \alpha_2 + \alpha_3 = 0$$

$$2\alpha_1 + \alpha_2 + 0 = 0$$

$$3\alpha_1 + 0 + \alpha_3 = 0.$$

On subtracting the first equation from the third we obtain

$$2\alpha_1 - \alpha_2 = 0$$

but

$$2\alpha_1 + \alpha_2 = 0, \text{hence } \alpha_1 = 0 = \alpha_2 \text{ and so } \alpha_3 = 0.$$

In this case the solution to the equation

$$\alpha_1\mathbf{x}_1 + \alpha_2\mathbf{x}_2 + \alpha_3\mathbf{x}_3 = \mathbf{0}$$

is trivial, and so $\mathbf{x}_1, \mathbf{x}_2$ and $\mathbf{x}_3$ are linearly independent.

Definition 2.10
A finite set of vectors $S = \{\mathbf{x}_1, \mathbf{x}_2, \ldots \mathbf{x}_n\}$ of a vector space V is said to be a **basis** for V if

(i) S spans (or generates) V, and

(ii) the vectors in S are linearly independent.
If the basis S consists of n vectors, we say that the vector space V is n-**dimensional**. This is denoted by dim $V = n$ (also see Theorem 2.6). The vectors in a basis S are sometimes referred to as the **basis vectors**.

As an example, if we consider the vectors $\mathbf{e}_1 = (1,0,0)$, $\mathbf{e}_2 = (0,1,0)$, $\mathbf{e}_3 = (0,0,1)$, then it is simple to verify that $S = \{\mathbf{e}_1, \mathbf{e}_2, \mathbf{e}_3\}$ satisfies conditions (i) and (ii) above and so is a basis for R^3. In fact it is called a **natural basis** for R^3. Of course dim $R^3 = 3$. $\mathbf{e}_1, \mathbf{e}_2, \mathbf{e}_3$ are called **natural** or **unit** vectors.

One of the important properties of a basis (which we could have used as an alternative definition for a basis) is summarised in the following theorem.

Theorem 2.2
If $S = \{\mathbf{x}_1, \mathbf{x}_2, \ldots \mathbf{x}_n\}$ is a basis of V, and $\mathbf{y}$ is any vector in V, then $\mathbf{y}$ can be written as a unique linear combination of the basis vectors.

Proof
Let us assume that

$$\mathbf{y} = a_1\mathbf{x}_1 + a_2\mathbf{x}_2 + \ldots + a_n\mathbf{x}_n \qquad \text{and}$$

$$\mathbf{y} = b_1\mathbf{x}_1 + b_2\mathbf{x}_2 + \ldots + b_n\mathbf{x}_n$$

where a_i and b_i are scalars.

On subtraction, we obtain:

$$\mathbf{0} = (a_1 - b_1)\mathbf{x}_1 + (a_2 - b_2)\mathbf{x}_2 + \ldots + (a_n - b_n)\mathbf{x}_n.$$

But (by definition) $\mathbf{x}_1, \mathbf{x}_2, \ldots \mathbf{x}_n$ are linearly independent, hence

$$a_i - b_i = 0$$

so that

$$a_i = b_i \qquad (i = 1, 2, \ldots n).$$

It will be useful to review briefly the concepts introduced when defining the generated space $L(S)$ and a basis (Def. 2.8 and 2.10). When defining $L(S)$ we consider a set of vectors $S \subseteq V$ which are not restricted so far as dependence or independence are concerned or indeed as to their number. In consequence $L(S)$ is a subspace of V which *could* be V itself.

On defining a basis for V we impose conditions on the vectors of the set S. They must be linearly independent, and the spanned space $L(S)$ *must* be V. Since the vectors in S must be linearly independent there is obviously a restriction on their number.

It is therefore clear that the set of basis vectors S is the smallest set of vectors which span the vector space $L(S) = V$.

A number of relevant questions arise. For example if $S = \{\mathbf{x}_1, \mathbf{x}_2, \ldots \mathbf{x}_n\}$ is a basis for V can we add another vector $\mathbf{w} \in V$ to this set so that the $(n+1)$ vectors $\mathbf{w}, \mathbf{x}_1, \ldots \mathbf{x}_n$ are linearly independent and span V?

If S is a basis for V, is it unique or could we choose another set of n linearly independent vectors in V to serve as a basis?

To answer these and other questions we need to prove a number of theorems.

In what follows we take α_i and β_i to be scalars.

Theorem 2.3

Let $\mathbf{x}_1, \mathbf{x}_2, \ldots \mathbf{x}_n$ be non-zero vectors $\in V$. These vectors are linearly dependent if and only if one of them, say $\mathbf{x}_i$, is a linear combination of the preceding vectors.

Proof

Assume that

$$\mathbf{x}_i = \alpha_1\mathbf{x}_1 + \alpha_2\mathbf{x}_2 + \ldots + \alpha_{i-1}\mathbf{x}_{i-1},$$

then

$$\alpha_1\mathbf{x}_1 + \alpha_2\mathbf{x}_2 + \ldots + \alpha_{i-1}\mathbf{x}_{i-1} + (-1)\mathbf{x}_i + o\mathbf{x}_{i+1} + \ldots + o\mathbf{x}_n = \mathbf{0}$$

At least one of the coefficients above (-1) is non-zero, hence the vectors $\mathbf{x}_1, \mathbf{x}_2, \ldots \mathbf{x}_n$ are linearly dependent.

Conversely, consider that the vectors are linearly dependent, that is

$$\alpha_1\mathbf{x}_1 + \alpha_2\mathbf{x}_2 + \ldots + \alpha_n\mathbf{x}_n = \mathbf{0}$$

and not all the $\alpha_i = 0$.

Let r be the largest integer such that $\alpha_r \neq 0$, that is, $1 \leq r \leq n$ and

$$\alpha_1\mathbf{x}_1 + \alpha_2\mathbf{x}_2 + \ldots + \alpha_r\mathbf{x}_r + o\mathbf{x}_{r+1} + \ldots + o\mathbf{x}_n = \mathbf{0}.$$

If $r = 1$, then

$$\alpha_1 \mathbf{x} = \mathbf{O} \text{ and so } \alpha_1 = 0.$$

This is impossible because the vectors $\{\mathbf{x}_i\}$ are linearly dependent. Hence $r > 1$ and since

$$\alpha_1 \mathbf{x}_1 + \alpha_2 \mathbf{x}_2 + \ldots + \alpha_r \mathbf{x}_r = \mathbf{O},$$

it follows that

$$\mathbf{x}_r = -\frac{1}{\alpha_r}(\alpha_1 \mathbf{x}_1 + \alpha_2 \mathbf{x}_2 + \ldots + \alpha_{r-1} \mathbf{x}_{r-1}).$$

This proves that $\mathbf{x}_r$ is a linear combination of the preceding vectors.

Theorem 2.4
If $S = \{\mathbf{x}_1, \mathbf{x}_2, \ldots \mathbf{x}_n\}$ is a set of non-zero vectors spanning the vector space V, then a subset of S is a basis of V.

Proof
If S is linearly independent (that is, the vector elements of S are linearly independent) then S is a basis of V.

If S is linearly dependent there is a vector $\mathbf{x}_i$ which (by Theorem 2.3) can be written as the combination

$$\mathbf{x}_i = \alpha_1 \mathbf{x}_1 + \ldots + \alpha_{i-1} \mathbf{x}_{i-1}.$$

Let $\mathbf{x} \in V$, then

$$\mathbf{x} = \beta_1 \mathbf{x}_1 + \ldots + \beta_i \mathbf{x}_i + \ldots + \beta_n \mathbf{x}_n$$

$$= \beta_1 \mathbf{x}_1 + \ldots + \beta_i(\alpha_1 \mathbf{x}_1 + \ldots + \alpha_{i-1} \mathbf{x}_{i-1}) + \ldots + \beta_n \mathbf{x}_n$$

$$= \gamma_1 \mathbf{x}_1 + \ldots + \gamma_{i-1} \mathbf{x}_{i-1} + \gamma_{i+1} \mathbf{x}_{i+1} + \ldots + \gamma_n \mathbf{x}_n$$

where

$$\gamma_j = \beta_j + \beta_i \alpha_j \quad (j=1,2,\ldots i-1)$$

and

$$\gamma_j = \beta_j \quad (j=i, i+1, \ldots n).$$

Hence the set of vectors $S_1 = \{\mathbf{x}_1, \mathbf{x}_2, \ldots \mathbf{x}_{i-1}, \mathbf{x}_{i+1}, \ldots \mathbf{x}_n\}$ span the vector space V.

If S_1 is linearly independent it is a basis for V, otherwise we use the process described above to delete a vector from S_1.

Continuing in this manner, the process will eventually yield (since S is a finite set) a linearly independent set S_r which is a basis for V.

Example 2.6

$S = \{\mathbf{x}_1,\mathbf{x}_2,\mathbf{x}_3,\mathbf{x}_4,\mathbf{x}_5\}$ is a set of vectors $\in R^3$ where $\mathbf{x}_1 = (1,-2,-1)$, $\mathbf{x}_2 = (1,2,0)$, $\mathbf{x}_3 = (1,1,1)$, $\mathbf{x}_4 = (1,-6,-2)$, and $\mathbf{x}_5 = (-1,0,-2)$.

Show that $\mathbf{x}_4$ is a linear combination of the preceding vectors, and find a subset of S which is a basis for R^3.

Solution

It is left as an exercise to show that

$$\mathbf{x}_4 = 2\mathbf{x}_1 - \mathbf{x}_2.$$

We consider the set $S_1 = \{\mathbf{x}_1,\mathbf{x}_2,\mathbf{x}_3,\mathbf{x}_5\}$, it is linearly dependent, indeed $\mathbf{x}_5 = \mathbf{x}_2 - 2\mathbf{x}_3$. On eliminating $\mathbf{x}_5$ we consider $S_2 = \{\mathbf{x}_1,\mathbf{x}_2,\mathbf{x}_3\}$. These three vectors are linearly independent and span R^3, hence S_2 is a basis for R^3.

Theorem 2.5

Let V be a vector space spanned by the set of vectors $S = \{\mathbf{x}_1,\mathbf{x}_2, \ldots \mathbf{x}_n\}$. Let $\{\mathbf{y}_1,\mathbf{y}_2, \ldots \mathbf{y}_r\}$ be a linearly independent set of vectors from V, then $n \geq r$.

Proof

Since the vectors in S span V we can write $\mathbf{y}_1 \in V$ as the linear combination

$$\mathbf{y}_1 = \alpha_1\mathbf{x}_1 + \alpha_2\mathbf{x}_2 + \ldots + \alpha_n\mathbf{x}_n$$

so that $S_1 = \{\mathbf{y}_1,\mathbf{x}_1,\mathbf{x}_2, \ldots \mathbf{x}_n\}$ is linearly dependent and also spans V.

It follows by Theorem 2.3 that one of the vectors, say $\mathbf{x}_r$ (note that it cannot be $\mathbf{y}_1$) is a linear combination of the preceding vectors, thus $Q_1 = \{\mathbf{y}_1,\mathbf{x}_1, \ldots \mathbf{x}_{r-1}, \mathbf{x}_{r+1}, \ldots \mathbf{x}_n\}$ spans V.

We next repeat the process and write $\mathbf{y}_2$ as a linear combination of vectors in Q_1 so that

$$S_2 = \{\mathbf{y}_2,\mathbf{y}_1,\mathbf{x}_1, \ldots \mathbf{x}_{r-1},\mathbf{x}_{r+1}, \ldots \mathbf{x}_n\}$$

is linearly dependent and spans V.

Again, one of the vectors from S_2, say $\mathbf{x}_j$, is a linear combination of the preceding ones. Note that $\mathbf{x}_j$ cannot be the first vector (by Theorem 2.3), nor can it be the second vector, since $\mathbf{y}_1$ is linearly independent of $\mathbf{y}_2$.

If we repeat this process r times and *assume that* $n \geq r$ we end up with the following linearly dependent set which spans V:

$$S_r = \{\mathbf{y}_r, \mathbf{y}_{r-1}, \ldots \mathbf{y}_1, \mathbf{x}_{k_1}, \ldots \mathbf{x}_{k_{n-r}}\}$$

where

$$\mathbf{x}_{k_1}, \ldots \mathbf{x}_{k_{n-r}} \text{ are some } (n-r) \text{ vectors from the set } S.$$

If, on the other hand, we *assume that* $n < r$ then we end up with the linearly dependent set

$$S_n = \{\mathbf{y}_n, \mathbf{y}_{n-1}, \ldots \mathbf{y}_1\} \qquad (n < r).$$

But S_n is a subset of $\{\mathbf{y}_1, \mathbf{y}_2, \ldots \mathbf{y}_r\}$ which is linearly independent. So our conclusion that S_n is linearly dependent is false, and our assumption that $n < r$ is wrong.

It follows that $n \geq r$.

We now prove the important theorem that every basis of a vector space V has the same number of vectors.

Theorem 2.6
If $S = \{\mathbf{x}_1, \mathbf{x}_2, \ldots \mathbf{x}_m\}$ and $Q = \{\mathbf{y}_1, \mathbf{y}_2, \ldots \mathbf{y}_n\}$ are bases for the vector space V, then $m = n$.

Proof
We use Theorem 2.5.

Since the vectors in S span V and the vectors in Q are linearly independent, $m \geq n$. Similarly $n \geq m$. It follows that $m = n$.

We are now in a position to justify Def. 2.10 on the dimension of a vector space. Indeed, although there are many possible bases for a vector space V, they all have the same number of elements, so it makes sense to define that number as *the* dimension of the vector space.

Example 2.7
Show that the set $S = \{2t^2, 2t-1, t+2\}$ is a basis for P_2 (all polynomial functions of degree 2 or less) so that dim $P_2 = 3$.

Solution
We must show that (i) S spans P_2 and (ii) S is linearly independent.

(i) Let $at^2 + bt + c$ be any polynomial in P_2, we must determine α,β,γ so that

$$at^2 + bt + c = \alpha(2t^2) + \beta(2t-1) + \gamma(t+2)$$

$$= 2\alpha t^2 + (2\beta+\gamma)t + (2\gamma-\beta),$$

that is,

$$a = 2\alpha$$

$$b = 2\beta + \gamma$$

$$c = 2\gamma - \beta\text{, hence } \alpha = \frac{1}{2}a, \beta = \frac{1}{5}(2b-c) \text{ and } \gamma = \frac{1}{5}(b+2c)$$

so S spans P_2.

(ii) Consider the equation

$$\alpha(2t^2) + \beta(2t-1) + \gamma(t+2) = 0$$

that is,

$$2\alpha t^2 + (2\beta+\gamma)t + (2\gamma-\beta) = 0.$$

Since this equation must hold for all values of t, it follows that

$$2\alpha = 0, 2\beta + \gamma = 0, \text{ and } 2\gamma - \beta = 0,$$

that is,

$$\alpha = \beta = \gamma = 0$$

so S is linearly independent. Hence S is a basis for P_2, and since there are 3 vectors in the basis

$$\dim P_2 = 3.$$

Example 2.8
V is a n-dimensional vector space and U is a subspace of V. Show that the dimension of $U \leq n$. If $\dim U = n$, then $U = V$.

Solution
Since elements of U are elements of V, the vectors of a basis of U are a set of linearly independent vectors of V.

Since dim $V = n$, it follows (as a consequence of Theorem 2.5) that any $(n+1)$ vectors in V are linearly dependent, and the maximum number of linearly independent vectors in V is n.

From the above two facts it follows that dim $U \leq n$.

If dim $U = n$ then a set of linearly independent vectors in U, say $S = \{\mathbf{x}_1, \mathbf{x}_2, \ldots \mathbf{x}_n\}$ is a basis of U. But S also belongs to V, hence (since dim $V = n$) it serves as a basis of V. Thus $U = V$.

To end this section we prove that any set of linearly independent elements in a vector space V can be used as part of a basis of V.

Theorem 2.7

V is a vector space such that dim $V = k$. If $S = \{\mathbf{x}_1, \mathbf{x}_2, \ldots \mathbf{x}_r\}$ be a set of linearly independent vectors in V, then S is a part of a basis of V.

Proof

Obviously $r \leq k$.

If $r = k$, then S is a basis of V. If $r < k$, then there exists a vector $\mathbf{y}_1 \in V$ such that $\mathbf{y}_1 \notin L(S)$.

Consider the equation

$$\alpha_1\mathbf{x}_1 + \alpha_2\mathbf{x}_2 + \ldots + \alpha_r\mathbf{x}_r + \alpha_{r+1}\mathbf{y}_1 = \mathbf{0}.$$

We find $\alpha_{r+1} = 0$ (since $\mathbf{y}_1 \notin L(S)$), hence

$$\alpha_1\mathbf{x}_1 + \alpha_2\mathbf{x}_2 + \ldots + \alpha_r\mathbf{x}_r = \mathbf{0}.$$

Since $\mathbf{x}_1, \mathbf{x}_2, \ldots \mathbf{x}_r$ are linearly independent

$$\alpha_1 = \alpha_2 = \ldots = \alpha_r = 0.$$

It follows that $\mathbf{x}_1, \mathbf{x}_2, \ldots \mathbf{x}_r$, $\mathbf{y}_1$ are linearly independent. We can repeat this process until we obtain k linearly independent vectors $\mathbf{x}_1, \mathbf{x}_2, \ldots \mathbf{x}_r, \mathbf{y}_1, \ldots \mathbf{y}_{k-r}$. These k vectors are a basis of V, and so the theorem is proved.

2.3 COORDINATES AND THE TRANSITION MATRIX

When defining a vector space Def. 2.10) we considered a linearly independent set of vectors $\{\mathbf{x}_1, \mathbf{x}_2, \ldots \mathbf{x}_n\}$. The order of the elements within this set was immaterial. We shall now consider an **ordered basis** $S = \{\mathbf{x}_1, \mathbf{x}_2, \ldots \mathbf{x}_n\}$ for V, so that for example, the set $\{\mathbf{x}_n, \mathbf{x}_1, \mathbf{x}_2, \ldots \mathbf{x}_{n-1}\}$ is a different ordered basis for V.

Definition 2.11
Let $S = \{\mathbf{x}_1, \mathbf{x}_2, \ldots \mathbf{x}_n\}$ be an ordered basis for an n-dimensional vector space V. Let $\mathbf{y}$ be any vector in V, then $\mathbf{y} = a_1\mathbf{x}_1 + a_2\mathbf{x}_2 + \ldots + a_n x_n$, for some scalars $a_1, a_2, \ldots a_n$. We call $[\mathbf{y}]_s = [a_1, a_2, \ldots a_n]'$ the **coordinate matrix of y relative to the ordered basis** S. The elements of $[\mathbf{y}]_s$ are called the coordinates of $\mathbf{y}$ with respect to S.

It is easy to verify that the coordinates of $\mathbf{y}$ relative to S are unique. Indeed, assume that we also have

$$\mathbf{y} = b_1\mathbf{x}_1 + b_2\mathbf{x}_2 + \ldots + b_n\mathbf{x}_n,$$

then, on subtracting

$$\mathbf{0} = (a_1 - b_1)\mathbf{x}_1 + (a_2 - b_2)\mathbf{x}_2 + \ldots + (a_n - b_n)\mathbf{x}_n.$$

But since the $\mathbf{x}_i$ $(i=1, 2, \ldots n)$ are linearly independent, it follows that

$$a_i - b_i = 0 \text{ for all } i.$$

Next we consider two vectors, $\mathbf{x}$ and $\mathbf{y} \in V$.

Let $\mathbf{x} = \sum_{i=1}^{n} a_i\mathbf{x}_i$ and $\mathbf{y} = \sum_{i=1}^{n} b_i\mathbf{x}_i$, then

$$\mathbf{x} + \mathbf{y} = \sum_{i=1}^{n} (a_i + b_i)\mathbf{x}_i.$$

It follows that the ith coordinate in the ordered basis S of $(\mathbf{x}+\mathbf{y})$ is (a_i+b_i). Also the ith coordinate of $c\mathbf{x}$ (c being a scalar) is ca_i.

It is clear that if we consider another basis for V, say $Q = \{\mathbf{y}_1, \mathbf{y}_2, \ldots y_n\}$, the coordinates of a vector $\mathbf{w} \in V$ with respect to S are different from the ones with respect to Q. The interesting question which we shall now attempt to answer is how the two coordinates of $\mathbf{w}$ are related. Assume that

$$\mathbf{w} = a_1\mathbf{x}_1 + a_2\mathbf{x}_2 + \ldots + a_n\mathbf{x}_n \tag{2.1}$$

and

$$\mathbf{w} = b_1\mathbf{y}_1 + b_2\mathbf{y}_2 + \ldots + b_n y_n \tag{2.2}$$

so that

$$[\mathbf{w}]_S = \begin{bmatrix} a_1 \\ a_2 \\ \vdots \\ a_n \end{bmatrix} \quad \text{and} \quad [\mathbf{w}]_Q = \begin{bmatrix} b_1 \\ b_2 \\ \vdots \\ b_n \end{bmatrix}.$$

Since $S = \{\mathbf{x}_1, \mathbf{x}_2, \ldots \mathbf{x}_n\}$ is a basis for V, and $\mathbf{y}_i (i = 1, 2, \ldots n)$ are vectors in V, there exist unique scalars p_{ij} such that

$$\begin{aligned} \mathbf{y}_1 &= p_{11}\mathbf{x}_1 + p_{21}\mathbf{x}_2 + \ldots + p_{n1}\mathbf{x}_n \\ \mathbf{y}_2 &= p_{12}\mathbf{x}_1 + p_{22}\mathbf{x}_2 + \ldots + p_{n2}\mathbf{x}_n \\ &\vdots \\ \mathbf{y}_n &= p_{1n}\mathbf{x}_1 + p_{2n}\mathbf{x}_2 + \ldots + p_{nn}\mathbf{x}_n, \end{aligned}$$

that is, the coordinates of $\mathbf{y}_i$ with respect to the ordered basis S are

$$(p_{1i}, p_{2i}, \ldots p_{ni}) \quad (i = 1, 2, \ldots n).$$

From (2.2) and (2.3) we obtain

$$\begin{aligned} \mathbf{w} &= b_1\mathbf{y}_1 + b_2\mathbf{y}_2 + \ldots + b_n\mathbf{y}_n \\ &= b_1(p_{11}\mathbf{x}_1 + p_{21}\mathbf{x}_2 + \ldots + p_{n1}\mathbf{x}_n) \\ &+ b_2(p_{12}\mathbf{x}_1 + p_{22}\mathbf{x}_2 + \ldots + p_{n2}\mathbf{x}_n) \\ &\vdots \\ &+ b_n(p_{1n}\mathbf{x}_1 + p_{2n}\mathbf{x}_2 + \ldots + p_{nn}\mathbf{x}_n) \\ &= (b_1p_{11} + b_2p_{12} + \ldots + b_np_{1n})\mathbf{x}_1 \\ &+ (b_1p_{21} + b_2p_{22} + \ldots + b_np_{2n})\mathbf{x}_2 \\ &\vdots \\ &+ (b_1p_{n1} + b_2p_{n2} + \ldots + b_np_{nn})\mathbf{x}_n. \end{aligned} \tag{2.4}$$

We have verified that the coordinates of a vector relative to an ordered basis are unique. It follows, in comparing (2.1) and (2.4), that

$$\begin{aligned} a_1 &= p_{11}b_1 + p_{12}b_2 + \ldots + p_{1n}b_n \\ a_2 &= p_{21}b_1 + p_{22}b_2 + \ldots + p_{2n}b_n \\ &\vdots \\ a_n &= p_{n1}b_1 + p_{n2}b_2 + \ldots + p_{nn}b_n, \end{aligned}$$

which can be written in matrix form as

$$[\mathbf{w}]_S = P[\mathbf{w}]_Q \tag{2.5}$$

where P is the $n \times n$ matrix whose $(i,j)^{\text{th}}$ element is p_{ij}.

It is a convention to write (2.5) using a simpler notation. We denote

$$[\mathbf{w}]_S \text{ by } X \text{ and } [\mathbf{w}]_Q \text{ by } Y, \text{ then (2.5) becomes } X = PY. \tag{2.6}$$

Definition 2.12

The **transition matrix** from the S-basis $\{\mathbf{x}_1,\mathbf{x}_2, \ldots \mathbf{x}_n\}$ to the Q-basis $\{\mathbf{y}_1,\mathbf{y}_2, \ldots \mathbf{y}_n\}$ is the matrix

$$P = [p_{ij}]$$

where the elements of the ith column are the coordinates of the basis vector $\mathbf{y}_i$ with respect to the basis S.

Example 2.9

Let $U = R^3, S = \left\{ \begin{bmatrix} 1 \\ 2 \\ -1 \end{bmatrix}, \begin{bmatrix} -1 \\ 0 \\ 0 \end{bmatrix}, \begin{bmatrix} 2 \\ 0 \\ -1 \end{bmatrix} \right\}$ and

$$Q = \left\{ \begin{bmatrix} 1 \\ 1 \\ 0 \end{bmatrix}, \begin{bmatrix} 2 \\ -1 \\ 0 \end{bmatrix}, \begin{bmatrix} 0 \\ 0 \\ 3 \end{bmatrix} \right\}.$$

(i) Find the transition matrix P from the S-basis to the Q-basis.

(ii) If $\mathbf{x} = \begin{bmatrix} 1 \\ -2 \\ 3 \end{bmatrix}$ is a vector relative to the natural basis in U, find $[\mathbf{x}]_S$, $[\mathbf{x}]_Q$ and verify the relation $[\mathbf{x}]_S = P[\mathbf{x}]_Q$.

Solution

(i) Since

$$\begin{bmatrix} 1 \\ 1 \\ 0 \end{bmatrix} = \frac{1}{2}\begin{bmatrix} 1 \\ 2 \\ -1 \end{bmatrix} - \frac{3}{2}\begin{bmatrix} -1 \\ 0 \\ 0 \end{bmatrix} - \frac{1}{2}\begin{bmatrix} 2 \\ 0 \\ -1 \end{bmatrix}$$

$$\begin{bmatrix} 2 \\ -1 \\ 0 \end{bmatrix} = -\frac{1}{2}\begin{bmatrix} 1 \\ 2 \\ -1 \end{bmatrix} - \frac{3}{2}\begin{bmatrix} -1 \\ 0 \\ 0 \end{bmatrix} + \frac{1}{2}\begin{bmatrix} 2 \\ 0 \\ -1 \end{bmatrix}$$

and
$$\begin{bmatrix} 0 \\ 0 \\ 3 \end{bmatrix} = 0\begin{bmatrix} 1 \\ 2 \\ -1 \end{bmatrix} - 6\begin{bmatrix} -1 \\ 0 \\ 0 \end{bmatrix} - 3\begin{bmatrix} 2 \\ 0 \\ -1 \end{bmatrix}$$

hence
$$P = \frac{1}{2}\begin{bmatrix} 1 & -1 & 0 \\ -3 & -3 & -12 \\ -1 & 1 & -6 \end{bmatrix}.$$

(ii) As

$$\begin{bmatrix} 1 \\ -2 \\ 3 \end{bmatrix} = -1\begin{bmatrix} 1 \\ 2 \\ -1 \end{bmatrix} - 6\begin{bmatrix} -1 \\ 0 \\ 0 \end{bmatrix} - 2\begin{bmatrix} 2 \\ 0 \\ -1 \end{bmatrix} \text{ then } [\mathbf{x}]_S = \begin{bmatrix} -1 \\ -6 \\ -2 \end{bmatrix}$$

and
$$\begin{bmatrix} 1 \\ -2 \\ 3 \end{bmatrix} = -1\begin{bmatrix} 1 \\ 1 \\ 0 \end{bmatrix} + 1\begin{bmatrix} 2 \\ -1 \\ 0 \end{bmatrix} + 1\begin{bmatrix} 0 \\ 0 \\ 3 \end{bmatrix}, \text{ then } [\mathbf{x}]_Q = \begin{bmatrix} -1 \\ 1 \\ 1 \end{bmatrix}.$$

Also

$$P[\mathbf{x}]_Q = \frac{1}{2}\begin{bmatrix} 1 & -1 & 0 \\ -3 & -3 & -12 \\ -1 & 1 & -6 \end{bmatrix}\begin{bmatrix} -1 \\ 1 \\ 1 \end{bmatrix} = \begin{bmatrix} -1 \\ -6 \\ -2 \end{bmatrix} = [\mathbf{x}]_S.$$

Since the columns of P are the representations of the basis Q, they are linearly independent. We shall see (in Sec. 4.2) that this implies that the rank of P is n so that it is a non-singular matrix and so is invertible.

Since $X = PY$, it follows that

$$Y = P^{-1}X$$

or, using the coordinate notation (see 2.5)

$$[\mathbf{w}]_Q = P^{-1}[\mathbf{w}]_S$$

for every vector $\mathbf{w} \in U$.

This equation implies that P^{-1} is the transition matrix from the Q-basis to the S-basis.

Example 2.10

Using the data of Ex. 2.9 show that $[\mathbf{x}]_Q = P^{-1}[\mathbf{x}]_S$.

Solution

The problem is to determine P^{-1}. We can use one of the techniques discussed in Chapter 8. On the other hand since P^{-1} is the transition matrix from the Q-basis to the S-basis it can be found by the same technique as that used in Ex. 2.9.

Since

$$\begin{bmatrix} 1 \\ 2 \\ -1 \end{bmatrix} = \frac{5}{3}\begin{bmatrix} 1 \\ 1 \\ 0 \end{bmatrix} - \frac{1}{3}\begin{bmatrix} 2 \\ -1 \\ 0 \end{bmatrix} - \frac{1}{3}\begin{bmatrix} 0 \\ 0 \\ 3 \end{bmatrix}$$

$$\begin{bmatrix} -1 \\ 0 \\ 0 \end{bmatrix} = -\frac{1}{3}\begin{bmatrix} 1 \\ 1 \\ 0 \end{bmatrix} - \frac{1}{3}\begin{bmatrix} 2 \\ -1 \\ 0 \end{bmatrix} + 0\begin{bmatrix} 0 \\ 0 \\ 3 \end{bmatrix}$$

and

$$\begin{bmatrix} 2 \\ 0 \\ -1 \end{bmatrix} = \frac{2}{3}\begin{bmatrix} 1 \\ 1 \\ 0 \end{bmatrix} + \frac{2}{3}\begin{bmatrix} 2 \\ -1 \\ 0 \end{bmatrix} - \frac{1}{3}\begin{bmatrix} 0 \\ 0 \\ 3 \end{bmatrix}$$

hence

$$P^{-1} = \frac{1}{3}\begin{bmatrix} 5 & -1 & 2 \\ -1 & -1 & 2 \\ -1 & 0 & -1 \end{bmatrix}$$

This result can be easily checked, since we must have $PP^{-1} = I$.

Finally,

$$P^{-1}[\mathbf{x}]_S = \frac{1}{3}\begin{bmatrix} 5 & -1 & 2 \\ -1 & -1 & 2 \\ -1 & 0 & -1 \end{bmatrix}\begin{bmatrix} -1 \\ -6 \\ -2 \end{bmatrix} = \begin{bmatrix} -1 \\ 1 \\ 1 \end{bmatrix} = [\mathbf{x}]_Q.$$

PROBLEMS FOR CHAPTER 2

1) Indicate which of the following is a vector space over a field F. (F is not defined explicitly but can be considered as R).
   (a) {All functions which are continuous on the interval $[-a,a]$} (a is a scalar).
   (b) {All continuous functions in a set G such that $f(0) = 0$ whenever $f \in G$}.
   (c) {All continuous functions in a set G such that $f(0) = \alpha$ whenever $f \in G$} (α is a number).
   (d) {All polynomials of degree n or less}.
   (e) {All polynomials of degree n}.
   (f) The set of all triples $S = \{(x,y,z); z = x+y\}$.
   (g) The set of all triples $S = \{(x,y,z); z = x+y+1\}$.

2) (a) V is the vector space of polynomials P_2 over the field F of integers modulo 3, that is if $\mathbf{v} \in V$ it is of the form

$$ax^2 + bx + c \text{ where } a,b,c \in F.$$

   Given $\mathbf{v}_1(x) = 2x^2 + x + 2$, find
   (i) $\mathbf{v}_2(x)$ such that $\mathbf{v}_1(x) + \mathbf{v}_2(x) = 0 \in F$.
   (ii) $\alpha\mathbf{v}_1(x)$ where $\alpha = 2$.
   (iii) $\mathbf{v}_1(x) + \mathbf{v}_2(x)$ where $\mathbf{v}_2(x) = x^2 + 2x + 2$.
   (b) If U is the set of polynomials $\{ax^2 + bx + c\}$ where a,b,c are integers modulo 4, determine whether or not U is a vector space.

3) Form the indicated linear combinations:

   (i) $\mathbf{x}_1 - 2\mathbf{x}_2 + \mathbf{x}_3, \mathbf{x}_1,\mathbf{x}_2,\mathbf{x}_3 \in V = R^3$

   $\mathbf{x}_1 = (1,1,1), \mathbf{x}_2 = (2,0,-1)$ and $\mathbf{x}_3 = (-1,-2,3)$.

   (ii) $\mathbf{v}_1 + \mathbf{v}_2 - 2\mathbf{v}_3, \mathbf{v}_1,\mathbf{v}_2,\mathbf{v}_3 \in V = P_3$

   $\mathbf{v}_1 = 2x^2 - x + 1, \mathbf{v}_2 = x^2 + x - 1$ and $\mathbf{v}_3 = x^2 - 2x + 2$.

   (iii) $\mathbf{z}_1 + \mathbf{z}_2 - 2\mathbf{z}_3, \mathbf{z}_1,\mathbf{z}_2,\mathbf{z}_3 \in C$ (field of complex numbers)

   $\mathbf{z}_1 = 2 - i \quad \mathbf{z}_2 = 1 + 2i \quad \mathbf{z}_3 = -2 + 2i$.

4) Show that the vectors $\mathbf{x}_1 = (-1,1,2)$, $\mathbf{x}_2 = (0,1,1)$ and $\mathbf{x}_3 = (1,2,0)$ are linearly independent and express $\mathbf{x} = (-2,1,-1)$ as a linear combination of $\mathbf{x}_1,\mathbf{x}_2$ and $\mathbf{x}_3$.

5) Examine each of the following two sets for independence. If a set is dependent, select a maximum independent subset and express the remaining vectors as a linear combination of these.

(a) $\mathbf{x}_1 = (-2,1,-1)$
$\mathbf{x}_2 = (1,-3,2)$
$\mathbf{x}_3 = (0,-5,3)$
$\mathbf{x}_4 = (1,1,2)$

(b) $\mathbf{y}_1 = (-1,-2,1)$
$\mathbf{y}_2 = (2,1,-1)$
$\mathbf{y}_3 = (1,-4,1)$.

6) Show that the spaces spanned respectively by $P = \{\mathbf{x}_1,\mathbf{x}_2\}$ and $Q = \{\mathbf{y}_1,\mathbf{y}_2,\mathbf{y}_3\}$, where

$$\mathbf{x}_1 = (1,1,0), \mathbf{x}_2 = (2,1,-1) \text{ and } \mathbf{y}_1 = (3,2,-1), \mathbf{y}_2 = (4,3,-1), \mathbf{y}_3 = (11,8,-3)$$

are identical.

7) $\mathbf{x} = [1,-2,-1]'$ relative to the natural basis. Find the coordinates of $\mathbf{x}$ relative to the basis $Q = \{\mathbf{y}_1,\mathbf{y}_2,\mathbf{y}_3\}$ where $\mathbf{y}_1 = [0,1,1]'$, $\mathbf{y}_2 = [1,1,1]'$ and $\mathbf{y}_3 = [1,0,1]'$.

8) Let $S = \{[1,1,1]', [1,1,0]', [-1,0,0]'\}$

$$Q = \{[0,-1,2]', [0,0,-1]', [1,0,0]'\}.$$

(a) Find the transition matrix P from the S to the Q basis.
(b) If $\mathbf{x} = [-1,2,-1]'$ relative to the natural basis in R^3 find $[\mathbf{x}]_Q$ and $[\mathbf{x}]_S$.
(c) Find P^{-1}, the transition matrix from the Q to the S basis.

CHAPTER 3

Linear Transformations

3.1 HOMOMORPHISMS

Having discussed vector spaces, we can now discuss linear transformations from one space to another. We shall see that a matrix represents a linear transformation, and this fact will enable us to have a much deeper understanding of matrix operations. We shall use this fact later on to obtain a relatively simple matrix representation of a linear transformation by choosing appropriate bases in the two vector spaces.

Definition 3.1
Let U and V be vector spaces. A function $T:U \to V$ is called a *linear transformation* of U into V if for all $\mathbf{x}, \mathbf{y} \in U$ and all $a, b, c \in F$.
(i) $T(\mathbf{x}+\mathbf{y}) = T(\mathbf{x}) + T(\mathbf{y})$, and
(ii) $T(c\mathbf{x}) = cT(\mathbf{x})$.
The conditions (i) and (ii) can be combined into one condition:

$$T(a\mathbf{x} + b\mathbf{y}) = aT(\mathbf{x}) + bT(\mathbf{y}).$$

Linear transformations are also known as **homomorphisms** as they preserve structural operations in vector spaces.

Example 3.1
Show that the mapping T defined by

$$T(\alpha,\beta) = (\beta,\alpha).$$

is a linear transformation of R^2 into R^2.

Solution
Let $\mathbf{x} = (\alpha_1,\beta_1)$ and $\mathbf{y} = (\alpha_2,\beta_2)$, then $\mathbf{x}, \mathbf{y} \in R^2$ which is a vector space, hence

$$a\mathbf{x} + b\mathbf{y} = (a\alpha_1+b\alpha_2, a\beta_1+b\beta_2).$$

By definition of T and the above equation,

$$\begin{aligned} T(a\mathbf{x}+b\mathbf{y}) &= (a\beta_1+b\beta_2, a\alpha_1+b\alpha_2) \\ &= a(\beta_1,\alpha_1) + b(\beta_2,\alpha_2) \\ &= aT(\mathbf{x}) + bT(\mathbf{y}). \end{aligned}$$

Hence, by Def. 3.1, T is a linear transformation.

Example 3.2
Determine whether the following are linear transformations:

(i) $T: R^3 \to R$ defined by $T(\alpha,\beta,\gamma) = \alpha\beta\gamma$, and

(ii) $T: R^2 \to R^3$ defined by $T(\alpha,\beta) = (\alpha,1+\beta,\alpha+\beta)$.

Solution
(i) Let $\mathbf{x} = (1,2,3)$ and $\mathbf{y} = (2,3,4)$, then $\mathbf{x} + \mathbf{y} = (3,5,7)$. $T(\mathbf{x}) = 1.2.3 = 6$ and $T(\mathbf{y}) = 2.3.4 = 24$.
Also $T(\mathbf{x}+\mathbf{y}) = T(3,5,7) = 3.5.7 = 105$ and $T(\mathbf{x}) + T(\mathbf{y}) = 6 + 24 = 30$.
Since $T(\mathbf{x}+\mathbf{y}) \neq T(\mathbf{x}) + T(\mathbf{y})$, T is not linear.

(ii) Let $\mathbf{x} = (1,2)$, $\mathbf{y} = (3,4)$ then $\mathbf{x} + \mathbf{y} = (4,6)$.
Also $T(\mathbf{x}) = (1,3,3)$, $T(\mathbf{y}) = (3,5,7)$ so that $T(\mathbf{x}) + T(\mathbf{y}) = (4,8,10)$.
$T(\mathbf{x}+\mathbf{y}) = (4,7,10)$
Since $T(\mathbf{x}+\mathbf{y}) \neq T(\mathbf{x}) + T(\mathbf{y})$, T is not linear.

The second part of the above example could have been solved rapidly if we had noted the following:
By Def. 3.1 (ii)

$$T(c\mathbf{x}) = cT(\mathbf{x}).$$

Substituting for $c = 0$, we obtain:

$$T(\mathbf{O}) = \mathbf{O}$$

that is, the zero vector in U is mapped into the zero vector in V.

In the above example 3.2 (ii) we have

$$T(\mathbf{O}) = (0,1,0) \neq 0,0,0)$$

hence T is not a linear transformation.

Definition 3.2

(i) A linear transformation $T: U \to V$ is said to be **one-to-one** (or just **one-one**) if whenever $\mathbf{x} \neq \mathbf{y}$ it follows that $T(\mathbf{x}) \neq T(\mathbf{y})$.

(ii) If A is a subset of U, then the **image** of A, denoted by $T(A) = \{\mathbf{v}: \mathbf{v} = T(\mathbf{x})$ for $\mathbf{x} \in A\}$. In particular the image of the space U is called the **image** of T and is denoted by $Im(T)$.

(iii) If the image of the space U is the space V, that is, if $Im(T) = V$, the homomorphism T is said to be a mapping **onto** V.

Definition 3.3

If $T: U \to V$ and

(1) T is a linear transformation

(2) T is one-one, and

(3) T is onto, then

T is called an **isomorphism** of U onto V. U is said to be **isomorphic** to V.

Example 3.3

Consider the functions defined as follows:

(a) $f: A \to B$ where $A = \{2,3,5\}$, $B = \{a,b,c,d\}$, and $f(2) = a$, $f(3) = d$, and $f(5) = b$.

(b) $g: C \to D$ where $C = \{8,9,10\}$, $D = \{x,y,z\}$, and $g(8) = y, g(9) = z, g(10) = x$.

Determine whether the mappings are one-to-one and onto.

Solution

Since each element in A has a different image, f is one-one. g is also one-one.

$c \in B$ but is not the image of any element in A – hence f is not onto.

$g: C \to D$ is onto since each element in D is the image of an element in C.

Since an isomorphism $T: U \to V$ is a one-to-one mapping, it follows that for every $\mathbf{y} \in V$, there is a unique $\mathbf{x} \in U$ such that $T(\mathbf{x}) = \mathbf{y}$. Hence we can define an **inverse mapping T^{-1}** by

$$T^{-1}(\mathbf{y}) = \mathbf{x}.$$

It is not difficult to show when T is an isomorphism so is T^{-1}.

Example 3.4
$T: R^2 \to R^2$ is defined by $T(\alpha,\beta) = (3\alpha,3\beta)$.
Show that T defines an isomorphism and find T^{-1}.

Solution
To show that T is an isomorphism we must show that it is
(1) a linear transformation,
(2) one-one, and
(3) onto.

$$T(a(\alpha_1,\beta_1) + b(\alpha_2,\beta_2)) = T((a\alpha_1+b\alpha_2),(a\beta_1+b\beta_2))$$

$$= (3a\alpha_1+3b\alpha_2,3a\beta_1+3b\beta_2)$$

$$= a(3\alpha_1,3\beta_1) + b(3\alpha_2,3\beta_2)$$

$$= aT(\alpha_1,\beta_1) + bT(\alpha_2,\beta_2).$$

Hence T is a linear transformation.
Consider any element in R^2, say (x,y). To this element there corresponds a unique element $(\alpha,\beta) \in R^2$ such that

$$T(\alpha,\beta) = (x,y)$$

hence T is one-one and onto.
The unique element (α,β) such that

$$T(\alpha,\beta) = (x,y) \text{ is}$$

$$(\alpha,\beta) = (\frac{1}{3}x, \frac{1}{3}y)$$

hence $T^{-1}: (x,y) \to (\frac{1}{3}x, \frac{1}{3}y)$.

It is interesting to note that the set of all linear transformations from a vector space U to a vector space V, denoted by $L(U,V)$, is itself a vector space, as proved in the following theorem.

Theorem 3.1
Let U and V be vector spaces (over the same field F). Let T_1 and T_2 be linear transformations from U to V. The sum of linear transformations $(T_1 + T_2)$ is defined by

$$(T_1 + T_2)(\mathbf{x}) = T_1(\mathbf{x}) + T_2(\mathbf{x}) \qquad (\mathbf{x} \in U)$$

and the function αT_1 is defined by

$$\alpha T_1(\mathbf{x}) = \alpha(T_1(\mathbf{x})) \qquad (\alpha \text{ being a scalar}).$$

The set of linear transformations from U to V together with the two operations of addition and scalar multiplication defined above, is a vector space.

Proof
For $\mathbf{x}_1, \mathbf{x}_2 \in U$, using the above definitions, we have

$$\begin{aligned}(T_1+T_2)(\alpha\mathbf{x}_1+\mathbf{x}_2) &= T_1(\alpha\mathbf{x}_1+\mathbf{x}_2) + T_2(\alpha\mathbf{x}_1+\mathbf{x}_2)\\ &= \alpha T_1(\mathbf{x}_1) + T_1(\mathbf{x}_2) + \alpha T_2(\mathbf{x}_1) + T_2(\mathbf{x}_2)\\ &= \alpha[T_1(\mathbf{x}_1) + T_2(\mathbf{x}_1)] + T_1(\mathbf{x}_2) + T_2(\mathbf{x}_2)\\ &= \alpha(T_1+T_2)(\mathbf{x}_1) + (T_1+T_2)(\mathbf{x}_2).\end{aligned}$$

This shows that (T_1+T_2) is a linear transformation.

Similarly we show that αT_1 is a linear transformation. Indeed, if a is another scalar

$$\begin{aligned}(\alpha T_1)(a\mathbf{x}_1 + \mathbf{x}_2) &= \alpha[T_1(a\mathbf{x}_1 + \mathbf{x}_2)]\\ &= \alpha T_1(a\mathbf{x}_1) + \alpha T_1(\mathbf{x}_2)\\ &= a[\alpha T_1(\mathbf{x}_1)] + \alpha T_1(\mathbf{x}_2).\end{aligned}$$

The fact that the remaining conditions for a vector space (see Def. 2.5) hold can be verified. For example the zero element of $L(U,V)$ is the zero transformation which maps each $\mathbf{x} \in U$ into $\mathbf{O} \in V$.

Finally in this section we shall define the Composition and the Inverse (a generalisation of discussion in Sec. 3.1) of a Linear Transformation.

Definition 3.4
Let T_1 and T_2 be linear transformations $T_1 : U \to V$ and $T_2 : V \to W$, then for all $\mathbf{x} \in U$, the **composition** (T_2T_1) is defined by

$$(T_2T_1)(\mathbf{x}) = T_2(T_1(\mathbf{x})).$$

From the definition, it is not difficult to verify that (T_2T_1) is a linear transformation from U to W. (See Problem 5).

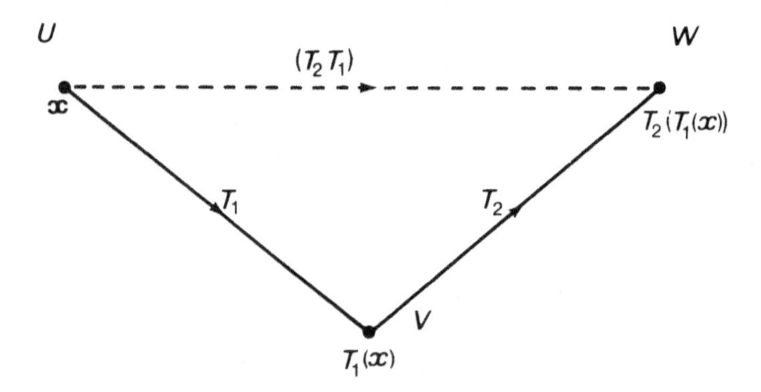

The **inverse** (if it exists – see Sec. 3.1) of a linear transformation $T_1 : U \to V$ is a linear transformation $T_2 : V \to U$ such that

$$(T_2T_1) = I_u \quad \text{and} \quad (T_1T_2) = I_v$$

where I_u, I_v are the identity transformations in U and V respectively.

3.2 ISOMORPHISM AND VECTOR SPACES

The importance of isomorphism in the context of vector spaces lies in the fact that the algebraic behaviour of two isomorphic vector spaces is identical although the elements of the vector spaces may be quite different. An interesting theorem follows:

Theorem 3.2
An n-dimensional vector space V is isomorphic to R^n.

Proof
Let $S = \{\mathbf{x}_1, \mathbf{x}_2, \ldots \mathbf{x}_n\}$ be an ordered basis for V. We define the mapping T as follows:

$$T : a_1\mathbf{x}_1 + a_2\mathbf{x}_2 + \ldots + a_n\mathbf{x}_n \to (a_1, a_2, \ldots . a_n).$$

Let $T(\mathbf{y}) = [\mathbf{y}]_s = (a_1, a_2, \ldots a_n) \in R^n$.
T is a mapping from V to R^n, and we now show that it is an isomorphism from V to R^n. First we show that T is one-one.
Let $[\mathbf{w}]_s = (b_1, b_2, \ldots b_n)$.

Suppose that $T(\mathbf{y}) = T(\mathbf{w})$ then $(a_1, a_2, \ldots a_n) = (b_1, b_2, \ldots b_n)$ so that $a_i = b_i$ $(i = 1, 2, \ldots n)$.
That is, $\mathbf{y} = \mathbf{w}$.

Next we show that T is onto.
Let $\mathbf{Z} = (c_1, c_2, \ldots c_n) \in R^n$, then there exists a vector $\mathbf{v} \in V$ such that $T(\mathbf{v}) = \mathbf{Z}$. Indeed let $c_1\mathbf{x}_1 + c_2\mathbf{x}_2 + \ldots + c_n\mathbf{x}_n = \mathbf{v}$, then the condition is satisfied.

Finally T is a linear transformation, since

$$T(\mathbf{y} + \mathbf{w}) = [\mathbf{y} + \mathbf{w}]_s = [\mathbf{y}]_s + [\mathbf{w}]_s = T(\mathbf{y}) + T(\mathbf{w}),$$

and

$$T(\alpha\mathbf{y}) = \alpha[\mathbf{y}]_s = \alpha T(\mathbf{y}).$$

It follows that T is an isomorphism from V to R^n.

Theorem 3.2 has many interesting consequences. In particular it serves as a unifying link for various types of vectors.

We have defined a vector $\mathbf{x}$ in an n-dimensional vector space V over the field F as

(i) the n-tuple of numbers $(x_1, x_2, \ldots x_n) \in F^n$.

Also we have seen, in Chapter 1, two particular n-tuples

(ii)

the one-column matrix $\begin{bmatrix} x_1 \\ x_2 \\ \vdots \\ x_n \end{bmatrix}$, and

(iii) the one-row matrix $[x_1, x_2, \ldots x_n]$.

Since these various representatives are all connected by isomorphisms, all calculations valid for one type of vectors are valid for all types. On the other hand we must be careful about how we associate vectors of different types. For example if $\mathbf{x}$ is a one-column matrix of order n and $\mathbf{y}$ is a one-row matrix of order n, we have no way of defining the sum $(\mathbf{x} + \mathbf{y})$. On the other hand, matrix multiplication is applicable and we find that

$\mathbf{x} \cdot \mathbf{y}$ is a matrix of order $n \times n$, whereas

$\mathbf{y} \cdot \mathbf{x}$ is the sum $(y_1x_1 + \ldots + y_nx_n)$.

Another interesting result of isomorphism $T : U \to V$ between two vector spaces U and V is that if $\{\mathbf{x}_1, \mathbf{x}_2, \ldots \mathbf{x}_r\}$ is a set of linearly independent vectors in U, then $\{T(\mathbf{x}_1), T(\mathbf{x}_2) \ldots T(\mathbf{x}_r)\}$ is a set of linearly independent vectors in V. In fact we generalise this in the following theorem:

Theorem 3.3
The mapping $T : U \to V$ is an isomorphism if and only if $\dim U = \dim V$.

Proof
If $\dim U = \dim V = n$, U and V are both isomorphic to R^n (Theorem 3.2) and so are isomorphic to each other.

Next consider that $T : U \to V$ is an isomorphism and that $S = \{\mathbf{x}_1, \mathbf{x}_2, \ldots \mathbf{x}_n\}$ is a basis for U, so that $\dim U = n$.

Let $Q = \{T(\mathbf{x}_1), T(\mathbf{x}_2), \ldots T(\mathbf{x}_n)\}$, we shall prove that Q is a basis for V. First to show that Q is a set of linearly independent vectors we consider

$$\alpha_1 T(\mathbf{x}_1) + \alpha_2 T(\mathbf{x}_2) + \ldots + \alpha_n T(\mathbf{x}_n) = \mathbf{0},$$

then

$$T(\alpha_1 \mathbf{x}_1 + \alpha_2 \mathbf{x}_2 + \ldots + \alpha_n \mathbf{x}_n) = \mathbf{0}$$

(since T is a linear transformation).

The zero vector on the right-hand side $\in V$, but since T is an isomorphism it is one-one, and maps the zero vector $\in U$ onto the zero vector $\in V$, $T(\mathbf{0}_u) = \mathbf{0}_v$, so that

$$\alpha_1 \mathbf{x}_1 + \alpha_2 \mathbf{x}_2 + \ldots + \alpha_n \mathbf{x}_n = \mathbf{0} \;(\in U)$$

But $\mathbf{x}_1, \mathbf{x}_2, \ldots \mathbf{x}_n$ are linearly independent, hence

$$\alpha_1 = \alpha_2 = \ldots = \alpha_n = 0$$

it follows that Q is linearly independent. Next we must show that Q spans V. Let $\mathbf{y}$ be any vector in V, then $\exists\, \mathbf{x} \in U$ such that

$$T(\mathbf{x}) = \mathbf{y}.$$

But $\mathbf{x} = \beta_1 \mathbf{x}_1 + \beta_2 \mathbf{x}_2 + \ldots + \beta_n \mathbf{x}_n$ (for some scalars $\beta_1, \beta_2, \ldots \beta_n$.) then

$$\begin{aligned} \mathbf{y} = T(\mathbf{x}) &= T(\beta_1 \mathbf{x}_1 + \beta_2 \mathbf{x}_2 + \ldots + \beta_n \mathbf{x}_n) \\ &= \beta_1 T(\mathbf{x}_1) + \beta_2 T(\mathbf{x}_2) + \ldots + \beta_n T(\mathbf{x}_n) \end{aligned}$$

so that Q spans V. This proves the theorem.

There is a particular isomorphism which is of great importance to us, we consider it in the next section.

3.3 LINEAR TRANSFORMATIONS AND MATRICES

We consider two vector spaces U and V of dimensions n and m respectively. Let $S = \{\mathbf{x}_1, \mathbf{x}_2, \ldots \mathbf{x}_n\}$ be an arbitrary ordered basis in U, and $Q = \{\mathbf{y}_1, \mathbf{y}_2, \ldots \mathbf{y}_m\}$ be an arbitrary ordered basis in V.

Let T be a linear transformation of U into V. Since $T(\mathbf{x}_1), T(\mathbf{x}_2), \ldots T(\mathbf{x}_n)$ are elements in V, they can be expressed uniquely in terms of the basis Q as follows

$$\begin{aligned} T(\mathbf{x}_1) &= a_{11}\mathbf{y}_1 + a_{21}\mathbf{y}_2 + \ldots + a_{m1}\mathbf{y}_m \\ T(\mathbf{x}_2) &= a_{12}\mathbf{y}_1 + a_{22}\mathbf{y}_2 + \ldots + a_{m2}\mathbf{y}_m \\ &\vdots \\ T(\mathbf{x}_n) &= a_{1n}\mathbf{y}_1 + a_{2n}\mathbf{y}_2 + \ldots + a_{mn}\mathbf{y}_m \end{aligned}$$

Note that the coordinate vector of $T(\mathbf{x}_j)$ with respect to the basis Q is

$$[T(\mathbf{x}_j)]_Q = (a_{1j}, a_{2j}, \ldots a_{mj}).$$

It will be useful to write the coordinate vector as the one-column matrix

$$[T(\mathbf{x}_j)]_Q = \begin{bmatrix} a_{1j} \\ a_{2j} \\ \vdots \\ a_{mj} \end{bmatrix}$$

Definition 3.5

The *matrix representing (or* **associated** *with) the linear transformation* T with respect to the bases S and Q is the matrix

$$A = [a_{ij}]$$

having as its jth column the coordinate vector $[T(\mathbf{x}_j)]_Q$ $(j = 1, 2, \ldots m)$.

Example 3.5
Let $T : R^3 \to R^2$ be defined by

$$T\left\{\begin{bmatrix} l \\ m \\ n \end{bmatrix}\right\} = \begin{bmatrix} l - m + 2n \\ 3l - 2m + n \end{bmatrix}.$$

$$\text{Let } S = \{\mathbf{x}_1, \mathbf{x}_2, \mathbf{x}_3\} = \left\{\begin{bmatrix} 1 \\ 0 \\ 1 \end{bmatrix}, \begin{bmatrix} 0 \\ 0 \\ 1 \end{bmatrix}, \begin{bmatrix} 0 \\ 1 \\ 1 \end{bmatrix}\right\}$$

$$\text{and } Q = \{\mathbf{y}_1, \mathbf{y}_2\} = \left\{\begin{bmatrix} 1 \\ 2 \end{bmatrix}, \begin{bmatrix} -1 \\ 1 \end{bmatrix}\right\}.$$

be ordered bases for R^3 and R^2 respectively.

Find the matrix representing T with respect to S and Q.

Solution
Notice that we can write the equation defining T as follows:

$$T\left\{\begin{bmatrix} l \\ m \\ n \end{bmatrix}\right\} = \begin{bmatrix} 1 & -1 & 2 \\ 3 & -2 & 1 \end{bmatrix}\begin{bmatrix} l \\ m \\ n \end{bmatrix}.$$

Now

$$T\left\{\begin{bmatrix} 1 \\ 0 \\ 1 \end{bmatrix}\right\} = \begin{bmatrix} 1 & -1 & 2 \\ 3 & -2 & 1 \end{bmatrix}\begin{bmatrix} 1 \\ 0 \\ 1 \end{bmatrix} = \begin{bmatrix} 3 \\ 4 \end{bmatrix}.$$

We must write $\begin{bmatrix} 3 \\ 4 \end{bmatrix}$ in terms of the basis Q, that is

$$\begin{bmatrix} 3 \\ 4 \end{bmatrix} = \alpha \begin{bmatrix} 1 \\ 2 \end{bmatrix} + \beta \begin{bmatrix} -1 \\ 1 \end{bmatrix} \Rightarrow \alpha = \frac{7}{3} \text{ and } \beta = -\frac{2}{3}, \text{ so that}$$

$$[T(\mathbf{x}_1)]_Q = \frac{1}{3} \begin{bmatrix} 7 \\ -2 \end{bmatrix}.$$

Similarly

$$T\left\{\begin{bmatrix} 0 \\ 0 \\ 1 \end{bmatrix}\right\} = \begin{bmatrix} 2 \\ 1 \end{bmatrix} = 1 \begin{bmatrix} 1 \\ 2 \end{bmatrix} -1 \begin{bmatrix} -1 \\ 1 \end{bmatrix} \text{ that is } [T(\mathbf{x}_2)]_Q = \begin{bmatrix} 1 \\ -1 \end{bmatrix}$$

$$\text{and } T\left\{\begin{bmatrix} 0 \\ 1 \\ 1 \end{bmatrix}\right\} = \begin{bmatrix} 1 \\ -1 \end{bmatrix} = 0 \begin{bmatrix} 1 \\ 2 \end{bmatrix} -1 \begin{bmatrix} -1 \\ 1 \end{bmatrix} \text{ that is } [T(\mathbf{x}_3)]_Q = \begin{bmatrix} 0 \\ -1 \end{bmatrix}.$$

It follows that the matrix representing T is

$$A = \begin{bmatrix} 7/3 & 1 & 0 \\ -2/3 & -1 & -1 \end{bmatrix}.$$

From the above discussion it follows that if

$$\mathbf{x} \in U \text{ and } T(\mathbf{x}) = \mathbf{y} \in V,$$

then

$$A[\mathbf{x}]_S = [\mathbf{y}]_Q.$$

We prove this statement below, but as an illustration we use the values of Ex. 3.5, and $\mathbf{x} = \begin{bmatrix} -1 \\ 2 \\ 1 \end{bmatrix}$.

We have $[\mathbf{x}]_S = \begin{bmatrix} -1 \\ 0 \\ 2 \end{bmatrix}$

and $A\,[\mathbf{x}]_S = \begin{bmatrix} 7/3 & 1 & 0 \\ -2/3 & -1 & -1 \end{bmatrix} \begin{bmatrix} -1 \\ 0 \\ 2 \end{bmatrix} = \begin{bmatrix} -7/3 \\ -4/3 \end{bmatrix}$.

Also $\mathbf{y} = T(\mathbf{x}) = \begin{bmatrix} 1 & -1 & 2 \\ 3 & -2 & 1 \end{bmatrix} \begin{bmatrix} -1 \\ 2 \\ 1 \end{bmatrix} = \begin{bmatrix} -1 \\ -6 \end{bmatrix}$

and $[\mathbf{y}]_Q = \begin{bmatrix} -7/3 \\ -4/3 \end{bmatrix}$.

We now prove this in general.

Theorem 3.4
Let $T : U \to V$ be a linear transformation where U is n-dimensional and V is m-dimensional. Let $S = \{\mathbf{x}_1, \mathbf{x}_2, \ldots \mathbf{x}_n\}$ and $Q = \{\mathbf{y}_1, \mathbf{y}_2, \ldots \mathbf{y}_m\}$ be ordered bases for U and V respectively. Let $A = [a_{ij}]$ be the matrix representing T with respect to S and Q, and $\mathbf{x} \in U$ and $T(\mathbf{x}) = \mathbf{y} \in V$, then $A[\mathbf{x}]_S = [\mathbf{y}]_Q$ where $[\mathbf{x}]_S$, $[\mathbf{y}]_Q$ are the coordinate vectors of $\mathbf{x}$ and $\mathbf{y}$ with respect to S and Q.

Proof
Since $\mathbf{x} \in U$, then

$$\mathbf{x} = \alpha_1 \mathbf{x}_1 + \alpha_2 \mathbf{x}_2 + \ldots + \alpha_n \mathbf{x}_n \text{ for some scalars } \alpha_1, \alpha_2, \ldots \alpha_n,$$

so that $[\mathbf{x}]_S = \begin{bmatrix} \alpha_1 \\ \alpha_2 \\ \vdots \\ \alpha_n \end{bmatrix}$.

Also $\mathbf{y} = T(\mathbf{x}) = \alpha_1 T(\mathbf{x}_1) + \alpha_2 T(\mathbf{x}_2) + \ldots + \alpha_n T(\mathbf{x}_n)$

$$= \alpha_1 (a_{11} \mathbf{y}_1 + a_{21} \mathbf{y}_2 + \ldots + a_{m1} \mathbf{y}_m)$$

$$+ \alpha_2 (a_{12} \mathbf{y}_1 + a_{22} \mathbf{y}_2 + \ldots + a_{m2} \mathbf{y}_m)$$

$$+ \ldots$$

$$+ \alpha_n (a_{1n} \mathbf{y}_1 + a_{2n} \mathbf{y}_2 + \ldots + a_{mn} \mathbf{y}_m)$$

$$= (\alpha_1 a_{11} + \alpha_2 a_{12} + \ldots + \alpha_n a_{1n}) \mathbf{y}_1 + (\alpha_1 a_{21} + \alpha_2 a_{22} + \ldots + \alpha_n a_{2n}) \mathbf{y}_2$$
$$+ \ldots + (\alpha_1 a_{m1} + \alpha_2 a_{m2} + \ldots + \alpha_n a_{mn}) \mathbf{y}_m. \quad (3.1)$$

Since $\mathbf{y} \in V$, there exist scalars $\beta_1, \beta_2, \ldots \beta_m$ such that

$$\mathbf{y} = \beta_1 \mathbf{y}_1 + \beta_2 \mathbf{y}_2 + \ldots + \beta_m \mathbf{y}_m. \quad (3.2)$$

so that $[\mathbf{y}]_Q = \begin{bmatrix} \beta_1 \\ \beta_2 \\ \vdots \\ \beta_m \end{bmatrix}$.

Equating the coefficients of $\mathbf{y}_i$ ($i = 1, \ldots m$) in (3.1) and (3.2), we obtain the following set of equations:

$$a_{11}\,\alpha_1 + a_{12}\,\alpha_2 + \ldots + a_{1n}\,\alpha_n = \beta_1$$

$$a_{21}\,\alpha_1 + a_{22}\,\alpha_2 + \ldots + a_{2n}\,\alpha_n = \beta_2$$

$$\vdots$$

$$a_{m1}\,\alpha_1 + a_{m2}\alpha_2 + \ldots + a_{mn}\,\alpha_n = \beta_m .$$

We can write this in matrix form as

$$\begin{bmatrix} a_{11} & a_{12} & \ldots & a_{1n} \\ a_{21} & a_{22} & \ldots & a_{2n} \\ \vdots & & & \\ a_{m1} & a_{m2} & \ldots & a_{mn} \end{bmatrix} \begin{bmatrix} \alpha_1 \\ \alpha_2 \\ \vdots \\ \alpha_n \end{bmatrix} = \begin{bmatrix} \beta_1 \\ \beta_2 \\ \vdots \\ \beta_m \end{bmatrix}$$

that is,

$$A\,[\mathbf{x}]_S = [\mathbf{y}]_Q .$$

Finally, it is not difficult to show that the matrix A as defined above is unique – the proof is left as an exercise.

Example 3.6

Let $A = \begin{bmatrix} 2 & 1 & -2 \\ 1 & -1 & 3 \end{bmatrix}$ be a given matrix.

Find the unique linear transformation $T : R^3 \to R^2$ whose representation with respect to S and Q is A, when

(i) S and Q are the natural bases for R^3 and R^2 respectively, and

(ii) $S = \left\{ \begin{bmatrix} 1 \\ 0 \\ 0 \end{bmatrix}, \begin{bmatrix} 1 \\ 1 \\ 0 \end{bmatrix}, \begin{bmatrix} 1 \\ 1 \\ 1 \end{bmatrix} \right\}$ and $Q = \left\{ \begin{bmatrix} 1 \\ 2 \end{bmatrix}, \begin{bmatrix} -2 \\ 3 \end{bmatrix} \right\}$.

Solution

(i) Let $\{\mathbf{E}_1, \mathbf{E}_2, \mathbf{E}_3\}$ be the natural basis for R^3 and $\{\mathbf{e}_1, \mathbf{e}_2\}$ be the natural basis for R^2.

By definition, the first column of A, that is, $\begin{bmatrix} 2 \\ 1 \end{bmatrix}$ is the coordinate vector of $T(\mathbf{E}_1)$ with respect to Q, so that

$$T(\mathbf{E}_1) = 2\mathbf{e}_1 + \mathbf{e}_2 = \begin{bmatrix} 2 \\ 1 \end{bmatrix}.$$

Similarly,

$$T(\mathbf{E}_2) = \begin{bmatrix} 1 \\ -1 \end{bmatrix} \text{ and } T(\mathbf{E}_3) = \begin{bmatrix} -2 \\ 3 \end{bmatrix}.$$

If $\mathbf{x} = \begin{bmatrix} \alpha \\ \beta \\ \gamma \end{bmatrix} \epsilon R^3$, then

$$T(\mathbf{x}) = T(\alpha\,\mathbf{E}_1 + \beta\,\mathbf{E}_2 + \gamma\,\mathbf{E}_3) = \alpha\,T(\mathbf{E}_1) + \beta\,T(\mathbf{E}_2) + \gamma\,T(\mathbf{E}_3)$$

$$= \begin{bmatrix} 2\alpha \\ \alpha \end{bmatrix} + \begin{bmatrix} \beta \\ -\beta \end{bmatrix} + \begin{bmatrix} -2\gamma \\ 3\gamma \end{bmatrix} = \begin{bmatrix} 2\alpha + \beta - 2\gamma \\ \alpha - \beta + 3\gamma \end{bmatrix} = \begin{bmatrix} 2 & 1 & -2 \\ 1 & -1 & 3 \end{bmatrix} \begin{bmatrix} \alpha \\ \beta \\ \gamma \end{bmatrix}$$

that is,

$$T(\mathbf{x}) = A\mathbf{x}.$$

so A is the matrix representation of T in this case.

(ii) Let $S = \{\mathbf{x}_1, \mathbf{x}_2, \mathbf{x}_3\}$ and $Q = \{\mathbf{y}_1, \mathbf{y}_2\}$.

By a similar argument to (i) above,

$$T(\mathbf{x}_1) = 2\,\mathbf{y}_1 + \mathbf{y}_2 = 2\begin{bmatrix} 1 \\ 2 \end{bmatrix} + \begin{bmatrix} -2 \\ 3 \end{bmatrix} = \begin{bmatrix} 0 \\ 7 \end{bmatrix}.$$

Similarly

$$T(\mathbf{x}_2) = \begin{bmatrix} 3 \\ -1 \end{bmatrix} \text{ and } T(\mathbf{x}_3) = \begin{bmatrix} -8 \\ 5 \end{bmatrix}.$$

This time,

$$T(\mathbf{x}) = \alpha T(\mathbf{x}_1) + \beta T(\mathbf{x}_2) + \gamma T(\mathbf{x}_3)$$

$$= \begin{bmatrix} 0 + 3\beta - 8\gamma \\ 7\alpha - \beta + 5\gamma \end{bmatrix} = \begin{bmatrix} 0 & 3 & -8 \\ 7 & -1 & 5 \end{bmatrix} \begin{bmatrix} \alpha \\ \beta \\ \gamma \end{bmatrix}.$$

In this case the matrix representation of T with respect to S and Q is

$$\begin{bmatrix} 0 & 3 & -8 \\ 7 & -1 & 5 \end{bmatrix}$$

and is different from the matrix A.

In this section it has become clear that there exists an intrinsic correspondence between matrices and linear transformations. This association is formalised in the following important theorem.

Theorem 3.5
Let $T : U \to V$ be a linear transformation where U is n-dimensional and V is m-dimensional (over F). Let A be the uniquely defined matrix representing T with respect to the pair of ordered bases S, Q for U and V respectively.

The function f mapping the space of all linear transformations from U to V, $L(U,V)$ to the space of all $m \times n$ matrices (over F) is an isomorphism.

Proof
Suppose that T_1 and $T_2 \in L(U,V)$, and that $A = [a_{ij}]$ and $B = [b_{ij}]$ are the matrices representing T_1 and T_2 with respect to the ordered bases S and Q.

Using the notation of Theorem 3.3 let

$$\mathbf{x} = \alpha_1\mathbf{x}_1 + \alpha_2\mathbf{x}_2 + \ldots + \alpha_n\mathbf{x}_n,$$

then

$$(T_1 + T_2)(\mathbf{x}) = T_1(\mathbf{x}) + T_2(\mathbf{x}).$$

The right-hand side can be written in the form of (3.1) as the sum

$$\sum_{i=1}^{m} (\sum_{j=1}^{n} a_{ij}\alpha_j)\mathbf{y}_i + \sum_{i=1}^{m} (\sum_{j=1}^{n} b_{ij}\alpha_j)\mathbf{y}_i$$

$$= \sum_{i=1}^{m} (\sum_{j=1}^{n} (a_{ij} + b_{ij})\alpha_j)\mathbf{y}_i.$$

This shows that the matrix representing $(T_1 + T_2)$ with respect to S and Q is the sum of the matrices A and B. By a similar method, we can show that the matrix representing αT_1 (where α is any scalar) is αA.

These two results imply that the function f is linear.

There is no difficulty in showing that f is one-one.

Finally, we must show that f is onto, that is, that any $m \times n$ matrix $A = [a_{ij}]$ represents a linear transformation of U to V.

Let $S = \{\mathbf{x}_1, \mathbf{x}_2, \ldots \mathbf{x}_n\}$ be a basis of U and $Q = \{\mathbf{y}_1, \mathbf{y}_2, \ldots \mathbf{y}_m\}$ be a basis of V. Using the elements of the jth column of A, we construct the vector

$$a_{ij}\mathbf{y}_1 + a_{2j}\mathbf{y}_2 + \ldots + a_{mj}\mathbf{y}_n \quad (j = 1, 2, \ldots n) \tag{3.3}$$

which is a uniquely defined element in the space V and so can be regarded as the image under some transformation T of the vector $\mathbf{x}_j \epsilon U$, that is, $T(\mathbf{x}_j)$ $(j = 1, 2, \ldots n)$.

Consider next any $\mathbf{x} \epsilon U$. It can be represented uniquely in the form

$$\mathbf{x} = \sum_{j=1}^{n} \alpha_j\mathbf{x}_j. \tag{3.4}$$

Since we wish to define T as a linear transformation, we must have

$$T(\mathbf{x}) = T(\sum_{j=1}^{n} \alpha_j\mathbf{x}_j)$$

$$= \sum_{j=1}^{n} \alpha_j T(\mathbf{x}_j) \quad \text{and using (3.3)}$$

$$= \sum_{j=1}^{n} \alpha_j \sum_{i=1}^{m} a_{ij} \mathbf{y}_i$$

$$= \sum_{i=1}^{m} \left(\sum_{j=1}^{n} a_{ij}\alpha_j \right) \mathbf{y}_i . \tag{3.5}$$

Hence, given any $\mathbf{x} \in U$, the scalars α_i are uniquely determined to write $\mathbf{x}$ as in (3.4). Next, given the matrix elements a_{ij}, all the coefficients in (3.5) are known, so that this equation uniquely defines the transformation T. It is but a matter of simple computation to verify that T is in fact a linear transformation. Since f is linear, one-one, and onto, it is an isomorphism and the theorem is proved.

The above theorem justifies our using results proved for linear transformations to matrices and conversely.

3.4 ORTHOGONAL TRANSFORMATIONS

Transformations of particular interest are those which preserve length – a property known as **isometry**.

A linear transformation T of a vector space V into itself having the isometry property is called:

orthogonal if V is over a real field
unitary if V is over a complex field.

Let T be a linear transformation of V over a field. Let $\mathbf{x} \in V$ and $\mathbf{y} = T(\mathbf{x})$, then T is an isometry if

$$|\mathbf{y}| = |\mathbf{x}|$$

or

$$|\mathbf{y}|^2 = |\mathbf{x}|^2$$

that is, if $\mathbf{y}'\mathbf{y} = \mathbf{x}'\mathbf{x}$.

Let the square matrix A be the representation of T (relative to the standard basis, say). Then

$\mathbf{y} = A\mathbf{x}$ and the above equation becomes

$$(A\mathbf{x})'(A\mathbf{x}) = \mathbf{x}'\mathbf{x}$$

that is,

$$\mathbf{x}'A'A\mathbf{x} - \mathbf{x}'\mathbf{x} = 0$$

or

$$x'(A'A-I)x = 0$$

so that

$$A'A = I.$$

Hence A is an *orthogonal* matrix (see Def. 1.16).

If the vector space V is over a complex field, a similar argument to the above shows that the square matrix A satisfies

$$A^*A = I$$

so that A is a *unitary* matrix (see Def. 1.16).

Transformations associated with rotations and reflexions are (in general) orthogonal.

For example, we can determine, using elementary trigonometry, the matrix associated with the rotation through an angle θ about the origin of a point $P(x_1,y_1)$ to $Q(x_2,y_2)$.

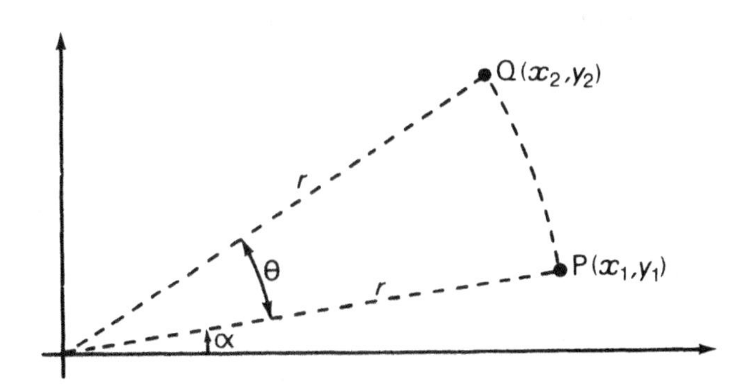

$$x_2 = r\cos(\alpha+\theta) = r\cos\alpha\cos\theta - r\sin\alpha\sin\theta$$

$$= x_1\cos\theta - y_1\sin\theta$$

$$y_2 = r\sin(\alpha+\theta) = r\sin\alpha\cos\theta + r\cos\alpha\sin\theta$$

$$= y_1\cos\theta + x_1\sin\theta.$$

Hence

$$\begin{bmatrix} x_2 \\ y_2 \end{bmatrix} = \begin{bmatrix} \cos\theta & -\sin\theta \\ \sin\theta & \cos\theta \end{bmatrix} \begin{bmatrix} x_1 \\ y_1 \end{bmatrix}.$$

The matrix

$$A = \begin{bmatrix} \cos\theta & -\sin\theta \\ \sin\theta & \cos\theta \end{bmatrix} \text{ is}$$

such that $AA' = I$, hence it is orthogonal.

Another example is that of a reflection in the y-axis, defined by

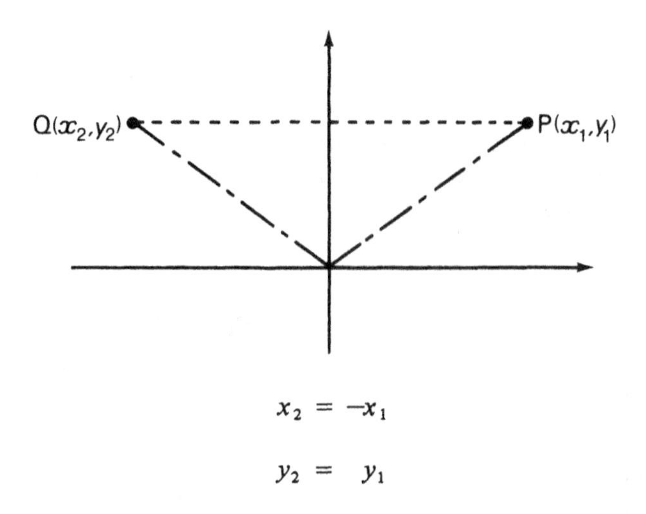

$$x_2 = -x_1$$

$$y_2 = y_1$$

that is,

$$\begin{bmatrix} x_2 \\ y_2 \end{bmatrix} = \begin{bmatrix} -1 & 0 \\ 0 & 1 \end{bmatrix} \begin{bmatrix} x_1 \\ y_1 \end{bmatrix} .$$

The matrix $A = \begin{bmatrix} -1 & 0 \\ 0 & 1 \end{bmatrix}$ is orthogonal.

Since an orthogonal transformation preserves length, it follows that a sequence of such transformations must also have this property; that is, the product of orthogonal matrices is itself an orthogonal matrix.

As an example we consider a rotation through an angle θ followed by a reflection in the y-axis. The matrix associated with this transformation is the product

$$\begin{bmatrix} -1 & 0 \\ 0 & 1 \end{bmatrix} \begin{bmatrix} \cos\theta & -\sin\theta \\ \sin\theta & \cos\theta \end{bmatrix}$$

$$= \begin{bmatrix} -\cos\theta & \sin\theta \\ \sin\theta & \cos\theta \end{bmatrix}$$

and is itself an orthogonal matrix.

3.5 GENERAL CHANGE OF BASES FOR A LINEAR TRANSFORMATION

In this section we shall consider the matrix representation of the linear transformation from U to V with respect to arbitrary bases, given the matrix representations of the transformation with respect to specified bases.

As usual we consider U and V to be vector spaces of dimensions n and m respectively. Let $S = \{\mathbf{x}_1, \mathbf{x}_2, \ldots \mathbf{x}_n\}$ and $S' = \{\mathbf{x}'_1, \mathbf{x}'_2, \ldots \mathbf{x}'_n\}$ be ordered bases for U, and $P = [p_{ij}]$ be the matrix of transition from S to S'. Let $Q = \{\mathbf{y}_1, \mathbf{y}_2, \ldots \mathbf{y}_m\}$ and $Q' = \{\mathbf{y}'_1, \mathbf{y}'_2, \ldots \mathbf{y}'_m\}$ be ordered bases for V, and H the matrix of transition from Q to Q'.

Let $T : U \to V$ be a linear transformation having the matrix representation $A = [a_{ij}]$ with respect to S and Q.

We shall first obtain the matrix $B = [b_{ij}]$ representing T with respect to S' and Q. In other words we shall consider how the matrix representing T changes when we change the basis S to S' in the domain of T.

Since T is represented by the matrix $A = [a_{ij}]$ with respect to S and Q, we have (see Def. 3.5)

$$T(\mathbf{x}_j) = \sum_{i=1}^{m} a_{ij}\, \mathbf{y}_i \quad (j = 1,2, \ldots n) \tag{3.6}$$

Since $P = [p_{ij}]$ is the matrix of transition from S to S', we have (see Def. 2.11 and Eqs. 2.3))

$$\mathbf{x}'_j = \sum_{k=1}^{n} p_{kj}\mathbf{x}_k \quad (j = 1,2, \ldots n). \tag{3.7}$$

By definition, $B = [b_{ij}]$ is the matrix representing T with respect to S' and Q, that is,

$$T(\mathbf{x}'_j) = \sum_{i=1}^{m} b_{ij}\, \mathbf{y}_j \quad (j = 1,2, \ldots m). \tag{3.8}$$

From (3.7)

$$T(\mathbf{x}'_j) = T(\sum_{k=1}^{n} p_{kj}\,\mathbf{x}_k)$$

$$= \sum_{k=1}^{n} p_{kj}\, T(\mathbf{x}_k),$$

and making use of (3.6)

$$= \sum_{k=1}^{n} p_{kj}(\sum_{i=1}^{m} a_{ik}\,\mathbf{y}_i)$$

$$= \sum_{i=1}^{m}\sum_{k=1}^{n} (a_{ik}\,p_{kj})\mathbf{y}_i. \qquad (3.9)$$

Since the right-hand sides of (3.8) and (3.9) represent the vector $T(\mathbf{x}_j)$, it follows that

$$b_{ij} = \sum_{k=1}^{n} a_{ik}\,p_{kj} \quad (i = 1,2,\ldots m, j = 1,2,\ldots n) \qquad (3.10)$$

and on expanding the right-hand side we recognise the $(i,j)^{\text{th}}$ element of the matrix AP. In matrix form, (3.10) is written as

$$B = AP \qquad (3.11)$$

which gives us the matrix B representing T with respect to S' and Q, given the matrix A representing T with respect to S and Q and the transition matrix P from S to S'.

Example 3.7
Let $T : U \to V$ be a linear transformation having the matrix representation

$$A = \begin{bmatrix} 3 & 5 \\ 1 & 2 \end{bmatrix}$$

with respect to $S = \{[1,1]',[-1,2]'\}$ and $Q = \{[-1,1]', [0,1]'$ where S and Q are bases for U and V respectively.

If $S' = \{[-1,1]', [2,-3]'\}$, find the matrix B representing T with respect to S' and Q.

Solution
We first need to find the matrix of transition P from S to S'.

$$[-1,1]' = p_{11}[1,1]' + p_{21}[-1,2]'$$

$$[2,-3]' = p_{12}[1,1]' + p_{22}[-1,2]'.$$

On solving these equations, we find

$$P = \frac{1}{3}\begin{bmatrix} -1 & 1 \\ 2 & -5 \end{bmatrix},$$

hence

$$B = AP = \frac{1}{3}\begin{bmatrix} 3 & 5 \\ 1 & 2 \end{bmatrix}\begin{bmatrix} -1 & 1 \\ 2 & -5 \end{bmatrix} = \frac{1}{3}\begin{bmatrix} 7 & -22 \\ 3 & -9 \end{bmatrix}.$$

For a clear understanding of the process carried out above, it is instructive to consider what happens to any vector in R^2, say $\mathbf{z} = [1,4]'$, under the above defined transformation.

Since $[1,4]' = 2[1,1]' + 1[-1,2]'$, it follows that $[\mathbf{z}]_S = [2,1]'$, and under the transformation defined by A

$$A[\mathbf{z}]_S = \begin{bmatrix} 3 & 5 \\ 1 & 2 \end{bmatrix}\begin{bmatrix} 2 \\ 1 \end{bmatrix} = \begin{bmatrix} 11 \\ 4 \end{bmatrix}_Q.$$

The coordinates of this vector are relative to the Q-basis, that is, they are $11[-1,1]' + 4[0,1]' = [-11,15]'$ relative to the standard basis for R^2.

On the other hand, since

$$[1,4]' = -11[-1,1]' - 5[2,-3]',$$

$$[\mathbf{z}]_{S'} = [-11,-5]'.$$

Hence under the transformation T, defined by the matrix B with respect to the bases S' and Q, we have

$$B[\mathbf{z}]_{S'} = \frac{1}{3}\begin{bmatrix} 7 & -22 \\ 3 & -9 \end{bmatrix}\begin{bmatrix} -11 \\ -5 \end{bmatrix} = \begin{bmatrix} 11 \\ 4 \end{bmatrix}_Q$$

which is of course the same vector as $A[\mathbf{z}]_S$.

The above discussion is theoretically interesting, but is of limited practical use. We shall see that when a change of bases is considered, it is usually done simultaneously both in the domain U and the co-domain V of T. Before discussing this general case, we shall consider a change of basis in V, from Q to Q'.

Since $H = [h_{ij}]$ is the matrix of transition from Q to Q' we have (similar to (3.7))

$$y'_j = \sum_{k=1}^{m} h_{kj}\mathbf{y}_k \quad (j = 1,2,\ldots m). \tag{3.12}$$

Let $C = [c_{ij}]$ be the matrix representing T with respect to S and Q'. We have

$$T(\mathbf{x}_j) = \sum_{i=1}^{m} c_{ij}\mathbf{y}'_i \quad (j = 1,2,\ldots n) \tag{3.13}$$

$$= \sum_{i=1}^{m} c_{ij} \left(\sum_{k=1}^{m} h_{ki}\mathbf{y}_k\right) \quad \text{(from (3.12))}$$

$$= \sum_{k=1}^{m} \left(\sum_{i=1}^{m} h_{ki}c_{ij}\right) \mathbf{y}_k. \tag{3.14}$$

Comparing Eqs. (3.6) and (3.14) and noting that the representation of the vector $T(\mathbf{x}_j)$ $(j = 1,2,\ldots n)$ is unique, it follows that

$$a_{kj} = \sum_{i=1}^{m} h_{ki}c_{ij} \quad (k = 1,2,\ldots m, j = 1,2,\ldots n). \tag{3.15}$$

In matrix form, (3.15) is written as

$$A = HC.$$

Hence the matrix C representing T with respect to S and Q', given the matrix representing T with respect to S and Q and the transition (hence non-singular) matrix H from Q to Q', is given by

$$C = H^{-1}A. \tag{3.16}$$

Example 3.8
The matrix representing $T : U \to V$ relative to the standard bases S and Q is

$$A = \begin{bmatrix} 3 & 5 \\ 1 & 2 \end{bmatrix}.$$

The basis for V is changed to $Q' = \{[3,1]', [5,2]'\}$. Find the matrix representing T relative to S and Q'.

Solution
The matrix of transition H from Q to Q' is

$$H = \begin{bmatrix} 3 & 5 \\ 1 & 2 \end{bmatrix}, \text{hence } H^{-1} = \begin{bmatrix} 2 & -5 \\ -1 & 3 \end{bmatrix}.$$

Thus $C = H^{-1}A = \begin{bmatrix} 1 & 0 \\ 0 & 1 \end{bmatrix}$.

We finally consider the general case, the representation of T when the bases in U and V are changed simultaneously.

Since Eqs. (3.11) and (3.16) give us the matrix representing T when a change in basis takes place in U and V respectively, on combining these two equations we obtain the matrix

$$H^{-1}AP \tag{3.17}$$

which represents T with respect to S' and Q', that is, with respect to a simultaneous change in bases in both U and V.

An interesting case of a simultaneous change of bases occurs when T is a linear transformation of a vector space U into itself, that is, when $U = V$ and when the bases S and Q are the same; that is, $S = Q$ and $S' = Q'$.

In these circumstances $H = P$, so that (3.17) becomes

$$P^{-1}AP. \tag{3.18}$$

Definition 3.6
Two matrices A and B are said to be **similar** if and only if there exists a non-singular matrix P such that

$$B = P^{-1}AP.$$

From the above discussion it is clear that two matrices are similar if they represent the same linear transformation relative (in general) to different bases.

Example 3.9
T is a linear transformation of a 2-dimensional vector space into itself having the matrix representation

$$A = \begin{bmatrix} 5 & -6 \\ 2 & -2 \end{bmatrix}$$

relative to the standard basis ($S = Q$).

Find the matrix representation of T relative to the basis $S' = \{[2,1]', [3,2]'\} = Q'$.

Solution
The matrix of transition from the standard to the S'-basis is

$$P = \begin{bmatrix} 2 & 3 \\ 1 & 2 \end{bmatrix}, \text{ so that } P^{-1} = \begin{bmatrix} 2 & -3 \\ -1 & 2 \end{bmatrix}.$$

It follows that the matrix of T relative to S' is

$$P^{-1}AP = \begin{bmatrix} 2 & -3 \\ -1 & 2 \end{bmatrix} \begin{bmatrix} 5 & -6 \\ 2 & -2 \end{bmatrix} \begin{bmatrix} 2 & 3 \\ 1 & 2 \end{bmatrix} = \begin{bmatrix} 2 & 0 \\ 0 & 1 \end{bmatrix}.$$

PROBLEMS FOR CHAPTER 3

1) Show that if P is a vector space of all polynomials, then

   (a) $D : P \to P$ where D is the differentiation operator,
   (b) $I : P \to P$ where I is the integration operator

   are linear transformations.

2) (a) Let D : $U \to V$ be a linear transformation (the 'derivative' of),

$U = P_4$ the set of polynomials of degree 4 and less

$V = P_3$ the set of polynomials of degree 3 and less.

Choosing the 'natural' bases for P_3 and P_4, that is, for P_3, $\{1,x,x^2,x^3\}$ and for P_4, $\{1,x,x^2,x^3,x^4\}$, find the matrix A representing D with respect to the two bases.

(b) Use A to find the derivative of $2x^4 - x^3 + x^2 - 2x + 1$.

3) (a) Let I : $U \to V$ be a linear transformation of a polynomial in P_2 to its integral, in P_3. Let the basis for P_2 be $S = \{\mathbf{z}_1,\mathbf{z}_2,\mathbf{z}_3\}$ and for P_3 be $Q = \{\mathbf{y}_1,\mathbf{y}_2,\mathbf{y}_3,\mathbf{y}_4\}$ where

$$\mathbf{z}_1 = 1, \mathbf{z}_2 = 1 + x, \mathbf{z}_3 = 1 - x + x^2,$$

$$\mathbf{y}_1 = 2, \mathbf{y}_2 = x - 1, \mathbf{y}_3 = x^2, \mathbf{y}_4 = x^3.$$

Find the matrix representing I with respect to the two bases.

4) Which of the following defines a linear transformation?

(i) $f(x,y) = (-y,x)$ (representing an anticlockwise rotation through 90°),
(ii) $f(x,y) = (-x,y)$ (representing a reflection in the y-axis),
(iii) $f(x,y) = (y,x)$ (representing a reflection in the line $y = x$);
(iv) $f(x,y) = (3x, 3y)$ (a magnification by 3);
(v) $f(x,y) = (2x-y, y+3x)$;
(vi) $f(x,y) = (2x+1, 2y-1)$.

In each case of a linear transformation determine the associated matrix.

5) (a) Given that f and g define two linear transformations T_1 and T_2, show that the composition (assumed defined for the two functions) $f \circ g$ also defines a linear transformation, $T = T_1T_2$.
(b) Relative to the natural basis, T_1 is represented by

$$A = \begin{bmatrix} a_{11} & a_{12} \\ a_{21} & a_{22} \end{bmatrix},$$

and T_2 by

$$B = \begin{bmatrix} b_{11} & b_{12} \\ b_{21} & b_{22} \end{bmatrix} .$$

Show that the matrix representing T_1T_2 is the product AB.

(c) f represents an anticlockwise rotation through $90°$, and g represents a reflection in the line $y = x$. Find the matrix corresponding to $f \circ g$.

6) $T : R^2 \to R^2$ is defined by

$$T(x,y) = (3x - 4y, 2x + y).$$

Find the matrix representing T relative to

(a) the natural basis;
(b) the basis in the domain is now chosen as $S_1 = \{(-1,1), (2,1)\}$, the co-domain basis is unchanged;
(c) the basis in the domain is the natural basis while the co-domain basis is $Q_1 = \{(1,1), (1,-1)\}$;
(d) the basis in the domain is S_1 and in the co-domain is Q_1.

CHAPTER 4

The Rank and the Determinant of a Matrix

4.1 THE KERNEL AND THE IMAGE SPACE OF A LINEAR TRANSFORMATION

There are two subspaces associated with a linear transformation T which are of particular interest. One is the image space of T already defined (Def. 3.2), the other is the kernel of T.

Definition 4.1
Consider the linear transformation $T : U \to V$.

(i) The **kernel** of T, denoted by $K(T)$ or ker T, is the subset of U, $\{\mathbf{x} \in U : T(\mathbf{x}) = \mathbf{O}_V\}$ where $\mathbf{O}_V$ is the zero vector in V. So the kernel of T is the subset of U consisting of all the vectors $\mathbf{x} \in U$ such that $T(\mathbf{x}) = \mathbf{O}_V$.

(ii) The **image** of T (or the **range** of T) denoted by Im(T) is the subset of $V = \{\mathbf{y} \in V : T(\mathbf{x}) = \mathbf{y}\}$ for all $\mathbf{x} \in U$. So the image of T is the subset of V consisting of vectors which are images, under T, of vectors in U.

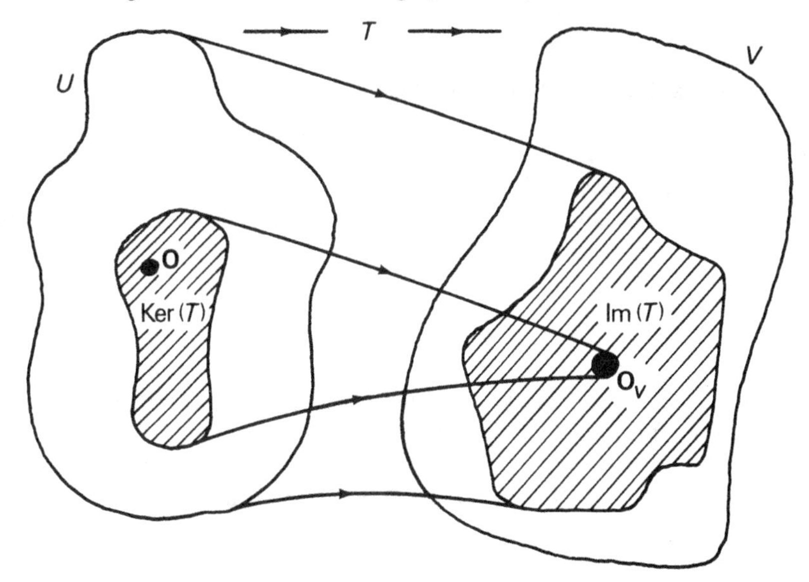

Example 4.1
Consider the linear transformation $T : R^3 \to R^2$ defined by

$$T([\alpha, \beta, \gamma]) = [\alpha, \alpha + \beta] \quad (\alpha,\beta,\gamma \in R).$$

Find (i) the kernel of T and (ii) the image space of T.

Solution
(i) To find the kernel of T we must find all vectors $\{\mathbf{x} = [\alpha,\beta,\gamma]\}$ in R^3 such that

$$T(\mathbf{x}) = [0,0].$$

When $T([\alpha,\beta,\gamma]) = [\alpha, \alpha + \beta] = [0,0]$
then $\alpha = 0$ and $\alpha + \beta = 0$, so that $\beta = 0$.
Hence the kernel of T consists of the set of vectors of the form

$$\ker T = \{[0,0,\gamma] : \gamma \in R\}.$$

(ii) Since α and β are any real numbers, the image vector $[\alpha,\alpha + \beta] \in R^2$ can be any vector in R^2. Hence the image of T is R^2.

We shall now prove that $\ker T$ is a subspace of U, and $\text{Im}(T)$ is a subspace of V.

Theorem 4.1
If $T : U \to V$ is a linear transformation, then

(i) $\ker T$ is a subspace of U.
(ii) $\text{Im}(T)$ is a subspace of V.
(iii) T is a one-one mapping, if and only if $\ker T = \{\mathbf{O}\}$, that is, if and only if the kernel of T consists only of the zero vector of U.

Proof

(i) Assume that $\mathbf{x}_1$ and $\mathbf{x}_2 \in \ker T$, then $T(\mathbf{x}_1) = \mathbf{O}$ and $T(\mathbf{x}_2) = \mathbf{O}$ ($\mathbf{O}$ is the zero vector in V).

Since $T(\mathbf{x}_1 + \mathbf{x}_2) = T(\mathbf{x}_1) + T(\mathbf{x}_2) = \mathbf{O} + \mathbf{O} = \mathbf{O}$, then

$$\mathbf{x}_1 + \mathbf{x}_2 \in \ker T.$$

Also

$$T(\alpha \mathbf{x}_1) = \alpha T(\mathbf{x}_1) = \mathbf{O},$$

so that

$$\alpha \mathbf{x}_1 \in \ker T \quad (\alpha \text{ being a scalar}).$$

Hence ker T is a subspace of U.

(ii) Let $\mathbf{y}_1$ and $\mathbf{y}_2 \in \text{Im}(T)$. There exist some vectors $\mathbf{x}_1, \mathbf{x}_2 \in U$, such that

$$T(\mathbf{x}_1) = \mathbf{y}_1 \text{ and } T(\mathbf{x}_2) = \mathbf{y}_2.$$

Since $\mathbf{x}_1, \mathbf{x}_2 \in U$, then $\mathbf{x}_1 + \mathbf{x}_2 \in U$, and by definition

$$T(\mathbf{x}_1 + \mathbf{x}_2) \in \text{Im}(T).$$

Now $\mathbf{y}_1 + \mathbf{y}_2 = T(\mathbf{x}_1) + T(\mathbf{x}_2) = T(\mathbf{x}_1 + \mathbf{x}_2)$, so that $\mathbf{y}_1 + \mathbf{y}_2 \in \text{Im}(T)$.
Similarly $\alpha\, \mathbf{y} \in \text{Im}(T)$ whenever $\mathbf{y} \in \text{Im}(T)$.
Hence $\text{Im}(T)$ is a subspace of V.

(iii) First assume that T is one-one.
Let $\mathbf{x} \in \ker T$, then

$$T(\mathbf{x}) = \mathbf{O}_v.$$

Since T is a linear transformation

$$T(\mathbf{O}_u) = \mathbf{O}_v.$$

It follows that $T(\mathbf{x}) = T(\mathbf{O}_u)$.
But T is one-one, so that $\mathbf{x} = \mathbf{O}_u$ hence

$$\ker T = \{\mathbf{O}_u\}$$

Next assume that $\ker T = \{\mathbf{O}_u\}$.
Let $\mathbf{x}_1, \mathbf{x}_2 \in U$ and assume that

$$T(\mathbf{x}_1) = T(\mathbf{x}_2),$$

then

$$T(\mathbf{x}_1) - T(\mathbf{x}_2) = T(\mathbf{x}_1 - \mathbf{x}_2) = \mathbf{O}_v,$$

that is,

$$\mathbf{x}_1 - \mathbf{x}_2 \in \ker T$$

so that

$$\mathbf{x}_1 - \mathbf{x}_2 = \mathbf{O}_u$$

that is,

$$\mathbf{x}_1 = \mathbf{x}_2.$$

It follows that T is one-one.

Example 4.2
$T : R^3 \to R^2$ is a linear transformation defined by

$$T([\alpha,\beta,\gamma]) = [\alpha + \beta, 2\alpha].$$

Find
(i) the basis and dimension of ker T, and
(ii) the basis and dimension of Im(T).

Solution
(i) $\ker T = \{\mathbf{x} : T(\mathbf{x}) = \mathbf{O}\}$, that is,
$\ker T = \{[\alpha,\beta,\gamma] : [\alpha + \beta, 2\alpha] = [0,0]\}$.

The above is true whenever $\alpha = 0 = \beta$.
Hence $\ker T = \{[0,0,\gamma], \gamma \in R\}$.
So the basis for ker T is a vector $[0,0,\gamma]$, and dim (ker T) = 1.

(ii) Since $T([\alpha,\beta,\gamma]) = [\alpha + \beta, 2\alpha]$
$= [\alpha, 2\alpha] + [\beta,0] = \alpha[1,2] + \beta[1,0]$,
it follows that $S = \{[1,2], [1,0]\}$ spans the space Im (T).
Since these two vectors are linearly independent, S is a basis for Im (T). It follows that dim Im $(T) = 2$.

The dimensions of ker T and Im (T) are important concepts in the development of matrix theory. They have special names.

Definition 4.2
(i) The dimension of kernel of T is known as the **nullity** of T.
(ii) The dimension of the image of T is known as the **rank** of T.

In the above example (4.2) the nullity of T is 1, the rank of T is 2. The sum of these two dimensions is equal to the dimension of the space U, in this case to dim $R^3 = 3$. This is in fact true in general, and the result is known as the **dimension theorem**.

Theorem 4.2 (the dimension theorem)
If U is an n-dimensional vector space and $T : U \to V$ is a linear transformation from U to V, then

$$n = \dim \text{Im}(T) + \dim \ker T. \tag{4.1}$$

Proof
By Theorem 4.1, ker T is a subspace of U, and it is therefore a vector space in its own right. Assume that $R = \{\mathbf{x}_1, \mathbf{x}_2, \ldots \mathbf{x}_k\}$ is a basis for ker T, so that dim ker $T = k$.

We can extend (Theorem 2.7) this set of linearly independent vectors to give a basis $S = \{\mathbf{x}_1, \mathbf{x}_2, \ldots \mathbf{x}_k, \mathbf{x}_{k+1} \ldots \mathbf{x}_n\}$ for U.

If we can show that the set $Q = \{T(\mathbf{x}_{k+1}), T(\mathbf{x}_{k+2}), \ldots T(\mathbf{x}_n)\}$ is a basis for Im (T), we shall have proved the theorem.

Let $\mathbf{x} \in U$, then for some scalars $\alpha_1, \ldots \alpha_n$,

$$\mathbf{x} = \alpha_1\mathbf{x}_1 + \ldots + \alpha_k\mathbf{x}_k + \alpha_{k+1}\mathbf{x}_{k+1} + \ldots + \alpha_n\mathbf{x}_n,$$

so that

$$\begin{aligned} T(\mathbf{x}) &= \alpha_1 T(\mathbf{x}_1) + \ldots + \alpha_k T(\mathbf{x}_k) + \alpha_{k+1} T(\mathbf{x}_{k+1}) + \ldots + \alpha_n T(\mathbf{x}_n) \\ &= \alpha_{k+1} T(\mathbf{x}_{k+1}) + \ldots + \alpha_n T(\mathbf{x}_n) \end{aligned}$$

since $\mathbf{x}_1, \mathbf{x}_2, \ldots \mathbf{x}_k \in \ker T$, that is,

$$T(\mathbf{x}_1) = T(\mathbf{x}_2) = \ldots = T(\mathbf{x}_k) = \mathbf{0}.$$

The above result demonstrates that the set Q spans Im (T).

Next we show that the vectors in Q are linearly independent. Assume that

$$\beta_{k+1} T(\mathbf{x}_{k+1}) + \beta_{k+2} T(\mathbf{x}_{k+2}) + \ldots + \beta_n T(\mathbf{x}_n) = \mathbf{0}$$

then

$$T(\beta_{k+1}\mathbf{x}_{k+1} + \beta_{k+2}\mathbf{x}_{k+2} + \ldots + \beta_n\mathbf{x}_n) = \mathbf{0},$$

so that

$$\beta_{k+1}\mathbf{x}_{k+1} + \beta_{k+2}\mathbf{x}_{k+2} + \ldots + \beta_n\mathbf{x}_n \in \ker T.$$

Since $R = \{\mathbf{x}_1, \mathbf{x}_2, \ldots \mathbf{x}_k\}$ is a basis for ker T, any vector in ker T is spanned by the basis vectors. This means that there exist scalars $b_1, b_2, \ldots b_k$ such that

$$\beta_{k+1}\mathbf{x}_{k+1} + \beta_{k+2}\mathbf{x}_{k+2} + \ldots + \beta_n \mathbf{x}_n = b_1\mathbf{x}_1 + b_2\mathbf{x}_2 + \ldots + b_k \mathbf{x}_k.$$

It follows that

$$b_1\mathbf{x}_1 + \ldots + b_k \mathbf{x}_k - \beta_{k+1}\mathbf{x}_{k+1} - \ldots - \beta_n \mathbf{x}_n = \mathbf{O}.$$

But the set S is linearly independent, hence

$$b_1 = \ldots = b_k = \beta_{k+1} = \ldots = \beta_n = 0,$$

and it follows that the set Q is linearly independent.

Having proved that Q both spans Im (T) and consists of linearly independent vectors, we have proved that it is a basis for Im (T).

Since the number of vectors in Q is $(n - k)$, the dim $Q = n - k$, that is,

$$\begin{aligned}\dim S = \dim U = n &= \dim R + \dim Q \\ &= \dim \ker T + \dim \operatorname{Im}(T).\end{aligned}$$

Example 4.3

Verify the dimension theorem for the linear transformation $T: R^3 \to R^2$ defined by

$$T([\alpha,\beta,\gamma]) = [\alpha + \beta, \beta + \gamma].$$

Solution

The ker T consists of vectors $\{\mathbf{x} = [\alpha,\beta,\gamma]\}$ such that

$$T([\mathbf{x}]) = [0,0],$$

that is,

$$\alpha + \beta = 0$$

$$\beta + \gamma = 0.$$

A possible nontrivial solution is $\alpha = \gamma = r, \beta = -r$.
Hence ker $T = \{[r, -r, r]; r \in R\}$,
and dim ker $T = 1$.

As $T([\alpha,\beta,\gamma]) = [\alpha + \beta, \beta + \gamma]$
$= [\alpha,\beta] + [\beta,\gamma]$,

so $S = \{[\alpha,\beta], [\beta,\gamma]\}$ spans Im (T). Since the two vectors are linearly independent, dim Im $(T) = 2$.

Also, since in this case $U = R^3$, dim $U = 3$, hence dim ker T + dim Im (T) $= 1 + 2 = 3 =$ dim U.

We shall discuss a number of applications of the Dimension Theorem in later chapters.

4.2 THE RANK OF A MATRIX

When considering a matrix A, we usually think of it as a rectangular array of numbers rather than as a representation of a linear transformation T. This in itself has no real disadvantage, since by Theorem 3.2 there is an isomorphism between the matrix A and the mapping T. On the other hand a more subtle problem arises involving the bases in U and V with respect to which A is the matrix representing T.

In general we consider a matrix A of order $m \times n$

$$A = \begin{bmatrix} a_{11} & a_{12} & \dots & a_{1n} \\ a_{21} & a_{22} & \dots & a_{2n} \\ \vdots & \vdots & & \\ a_{m1} & a_{m2} & \dots & a_{mn} \end{bmatrix} \tag{4.2}$$

which we know represents some linear transformation T. We are not given any further information, but since A is of order $m \times n$, we know that it is a matrix representing a mapping $T : U \to V$ where U has dimension n and V has dimension m. It follows that there is a basis $S = \{\mathbf{x}_1, \mathbf{x}_2, \dots \mathbf{x}_n\}$ in U, and a basis $Q = \{\mathbf{y}_1, \mathbf{y}_2, \dots \mathbf{y}_m\}$ in V, such that

$$T(\mathbf{x}_j) = \sum_{i=1}^{m} a_{ij}\mathbf{y}_i \quad (j = 1,2, \dots n). \tag{4.3}$$

So (see Def. 3.5) the coordinate vector $[T(\mathbf{x}_j)]_Q$ is the m-tuple $(a_{1j}, a_{2j}, \dots a_{mj})$, the jth column of A.

In conclusion we note that the columns of A represent respectively the vectors $T(\mathbf{x}_1)$, $T(\mathbf{x}_2)$, $\dots$ $T(\mathbf{x}_n)$ relative to the basis Q. It is instructive to visualise the above conclusions in matrix theory terms. As A is of order $m \times n$, Def. 4.1 can be written as

$$\text{Im}(T) = \{\mathbf{y} \in R^m : \mathbf{y} = A\mathbf{x} \text{ for all } \mathbf{x} \in R^n\}.$$

If $\mathbf{y} \in \text{Im}(T)$, then there exists some $\mathbf{x} \in R^n$ such that

$$\mathbf{y} = A\mathbf{x} = \begin{bmatrix} a_{11} & a_{12} & \dots & a_{1n} \\ a_{21} & a_{22} & \dots & a_{2n} \\ \vdots & & & \\ a_{m1} & a_{m2} & \dots & a_{mn} \end{bmatrix} \begin{bmatrix} x_1 \\ x_2 \\ \\ x_n \end{bmatrix}$$

$$= x_1 \begin{bmatrix} a_{11} \\ a_{21} \\ \vdots \\ a_{m1} \end{bmatrix} + x_2 \begin{bmatrix} a_{12} \\ a_{22} \\ \vdots \\ a_{m2} \end{bmatrix} + \dots + x_n \begin{bmatrix} a_{1n} \\ a_{2n} \\ \vdots \\ a_{mn} \end{bmatrix}. \tag{4.4}$$

This shows that $\mathbf{y}$ is a linear combination of the columns of A.

In the previous section (see Def. 4.2) we have defined the *rank* of T as the dimension of Im (T); we now prove the following theorem:

Theorem 4.3
Given $T : U \to V$

$$\text{Rank } (T) \leq \min \{n,m\}$$

where n is the dimension of U, and m is the dimension of V.

Proof
Let $S = \{\mathbf{x}_1, \mathbf{x}_2, \dots \mathbf{x}_n\}$ be a basis of U, then for any $\mathbf{x} \in U$, $\mathbf{x} = \sum_{i=1}^{n} \alpha_i \mathbf{x}_i$ where α_i are constants. Hence

$$T(\mathbf{x}) = \sum_{i=1}^{n} \alpha_i T(\mathbf{x}_i). \tag{4.5}$$

This equation implies that the image space of T, that is Im(T), is spanned by $\{T(\mathbf{x}_1), T(\mathbf{x}_2), \dots T(\mathbf{x}_n)\}$. Thus

$$\text{Rank}(T) = \dim \text{Im}(T) = \dim\{T(\mathbf{x}_1), T(\mathbf{x}_2), \ldots T(\mathbf{x}_n)\}.$$

$$\leq n.$$

Also by Def. 4.1, Im(T) is a subset of V, hence

$$\text{Rank}(T) = \dim \text{Im}(T) \leq \dim V = m.$$

We finally conclude that

$$\text{Rank}(T) \leq \min(n, m)$$

Corollary
If we consider any subspace W of U, then $\dim\{T(W)\} \leq \min\{\dim W, \dim V\}$. This corollary is obvious when we look at the proof of Theorem 4.3 and note that since W is a subspace of U it is spanned by a subset of S (which of course can be S itself.)

In the above discussion we noted that the set $\{T(\mathbf{x}_1), T(\mathbf{x}_2), \ldots T(\mathbf{x}_n)\}$ is isomorphic to the set of column vectors of A and spans the space Im(T). We conclude that the rank of T is equal to the maximum number of independent columns of A. This leads us to the following definition:

Definition 4.3
The **column rank** of A is the maximum number of linearly independent columns of A.

There is a similar definition for the row rank of A.

Definition 4.4
The **row rank** of A is the maximum number of linearly independent rows of A.

Theorem 4.4
The row rank of A is equal to the column rank of A.

Proof
Let A be a matrix of order $m \times n$ having a column rank equal to r.

A has the form of Eq. (4.2), and we can consider the columns of A to be column vectors of order m which we shall denote by $\mathbf{a}_1, \mathbf{a}_2, \ldots \mathbf{a}_n$.

Since the column rank of A is r, we know that exactly r of the vectors are linearly independent. There is no loss of generality for our purposes in proving this theorem to consider the first r vectors, that is, $\mathbf{a}_1, \mathbf{a}_2, \ldots \mathbf{a}_r$ to be linearly independent.

Every column of A can be expressed as a linear combination of these r vectors, for example the jth column can be written as

$$\begin{bmatrix} a_{1j} \\ a_{2j} \\ \vdots \\ a_{ij} \\ \vdots \\ a_{mj} \end{bmatrix} = h_{1j} \begin{bmatrix} a_{11} \\ a_{21} \\ \vdots \\ a_{i1} \\ \vdots \\ a_{m1} \end{bmatrix} + h_{2j} \begin{bmatrix} a_{12} \\ a_{22} \\ \vdots \\ a_{i2} \\ \vdots \\ a_{m2} \end{bmatrix} + \ldots + h_{rj} \begin{bmatrix} a_{1r} \\ a_{2r} \\ \vdots \\ a_{ir} \\ \vdots \\ a_{mr} \end{bmatrix} \quad (j = 1,2, \ldots n) \tag{4.6}$$

where h_{ij} $(i = 1, \ldots r; j = 1, \ldots n)$ are some constants. Comparing components in (4.6), we obtain:

$$a_{ij} = h_{1j}a_{i1} + h_{2j}a_{i2} + \ldots + h_{rj}a_{ir} \quad (i = 1, \ldots m, j = 1, \ldots n). \tag{4.7}$$

Considering (4.7) for $j = 1,2, \ldots n$ we obtain the equations

$$\begin{aligned} a_{i1} &= h_{11}a_{i1} + h_{21}a_{i2} + \ldots + h_{r1}a_{ir} \\ a_{i2} &= h_{12}a_{i1} + h_{22}a_{i2} + \ldots + h_{r2}a_{ir} \\ &\vdots \\ a_{in} &= h_{1n}a_{i1} + h_{2n}a_{i2} + \ldots + h_{rn}a_{ir} \quad (i = 1,2, \ldots m). \end{aligned} \tag{4.8}$$

We can write (4.8) in vector form as

$$[a_{i1}, a_{i2}, \ldots a_{in}] = a_{i1}\mathbf{h}_1 + a_{i2}\mathbf{h}_2 + \ldots + a_{ir}\mathbf{h}_r \quad (i = 1,2, \ldots m) \tag{4.9}$$

where

$$\mathbf{h}_k = [h_{k1}, h_{k2}, \ldots h_{kn}] \quad (k = 1,2, \ldots r).$$

But the left-hand side of (4.9) is the ith row vector of $A (i = 1,2, \ldots m)$. So every row vector of A is a linear combination of the r vectors $\mathbf{h}_1, \mathbf{h}_2, \ldots \mathbf{h}_r$. This implies that the subspace spanned by the rows of A has dimension at most r.

Hence

$$\text{row rank of } A \leq r = \text{column rank of } A. \tag{4.10}$$

Applying a similar argument to A', the transpose of A, we conclude

$$\text{row rank of } A' \leq \text{column rank of } A'. \tag{4.11}$$

But

$$\text{the columns of } A \text{ are the rows of } A'$$

and

$$\text{the rows of } A \text{ are the columns of } A'$$

so that (4.11) is analogous to

$$\text{column rank of } A \leq \text{row rank of } A. \tag{4.12}$$

Eqs (4.10) and (4.12) together imply that

$$r = \text{row rank of } A = \text{column rank of } A.$$

We call r the **rank of the matrix** A.

Notation We denote rank of A by $r(A)$.

Example 4.4
Find the rank of the matrix

$$A = \begin{bmatrix} 1 & 2 & 3 & 0 \\ 0 & -1 & -1 & 1 \\ -1 & 0 & -1 & -2 \end{bmatrix}.$$

Solution
On examining the column vectors of A, we notice that only two of them are independent, indeed we can write the third and fourth columns as a linear combination of the first two:

$$\begin{bmatrix} 3 \\ -1 \\ -1 \end{bmatrix} = \begin{bmatrix} 1 \\ 0 \\ -1 \end{bmatrix} + \begin{bmatrix} 2 \\ -1 \\ 0 \end{bmatrix} \text{ and } \begin{bmatrix} 0 \\ 1 \\ -2 \end{bmatrix} = 2\begin{bmatrix} 1 \\ 0 \\ -1 \end{bmatrix} - \begin{bmatrix} 2 \\ -1 \\ 0 \end{bmatrix}.$$

Hence the (column) rank of A is 2. We can verify that the row rank of A is also 2.

The first two rows vectors are obviously independent, and

$$[-1,0,-1,-2] = -[1,2,3,0] - 2[0,-1,-1,1].$$

Hence the (row) rank is 2.

Theorem 4.5

Given a matrix A of order $m \times n$ and of rank r, there exist non-singular matrices H and P such that

$$H^{-1}AP = \left[\begin{array}{c|c} I_r & O_{r,n-r} \\ \hline O_{m-r,r} & O_{m-r,n-r} \end{array}\right]$$

where H is of order $m \times m$, P is of order $n \times n$.

I_r is the unit matrix of order r, and the O matrices are zero matrices of order indicated.

Proof

A can be considered as the matrix representing a linear transformation $T: U \to V$ relative to standard bases S and Q. U and V are of dimension n and m respectively. By Theorem 4.2

$$\dim \ker(T) = n - \dim \operatorname{Im}(T) = n - r.$$

We can choose a basis $\{\mathbf{x}_{r+1}, \mathbf{x}_{r+2}, \ldots \mathbf{x}_n\}$ for ker (T) which (by Theorem 2.6) we can extend to form a basis

$$S' = \{\mathbf{x}_1, \ldots \mathbf{x}_r, \mathbf{x}_{r+1}, \ldots \mathbf{x}_n\} \text{ for } U.$$

We now have

$$T(\mathbf{x}_i) = \begin{cases} \mathbf{y}_i & (i = 1,2, \ldots r) \\ \mathbf{0} & (i = r+1, \ldots n). \end{cases} \tag{4.13}$$

We have shown in Theorem 4.2 that the set $\{\mathbf{y}_1, \mathbf{y}_2, \ldots \mathbf{y}_r\}$ is a basis for Im(T). We now extend this basis to $Q' = \{\mathbf{y}_1, \ldots \mathbf{y}_r, \mathbf{y}_{r+1}, \ldots \mathbf{y}_m\}$ for V. (4.13) shows that relative to the bases S' and Q' the matrix representing T has the required form

$$\left[\begin{array}{c|c} I_r & 0 \\ \hline 0 & 0 \end{array}\right].$$

Finally, from the discussion in Sec. 3.4, if P and H are the matrices of transition (hence non-singular) from S to S' and Q to Q' respectively, then the matrix representing T relative to S' and Q' is

$$H^{-1}AP$$

and has the above form.

Example 4.5
Find the matrices H and P such that the matrix A of Ex. 4.4 is transformed into the

$$\left[\begin{array}{c|c} I_2 & 0 \\ \hline 0 & 0 \end{array}\right]$$

form.

Solution
In Ex. 4.1we found that $r(A) = 2$, hence

$$\dim \ker(T) = n - r = 4 - 2 = 2.$$

It is not difficult to verify that

$$\{[1,4,-3,1]', [-2,1,0,1]'\}$$

is a basis for ker (T).

We extend this basis to a basis for U, for example

$$S' = \left\{ \begin{bmatrix} 1 \\ 0 \\ 0 \\ 0 \end{bmatrix}, \begin{bmatrix} 0 \\ 1 \\ 0 \\ 0 \end{bmatrix}, \begin{bmatrix} 1 \\ 4 \\ -3 \\ 1 \end{bmatrix}, \begin{bmatrix} -2 \\ 1 \\ 0 \\ 1 \end{bmatrix} \right\}.$$

The corresponding vectors $\mathbf{y}_i$ are

$$\mathbf{y}_1 = [1,0,-1]' \text{ and } \mathbf{y}_2 = [2,-1,0]'.$$

We can extend this to a basis for V, take for example

$$\mathbf{y}_3 = [1,0,0]', \text{ so that } Q' = \{\mathbf{y}_1,\mathbf{y}_2,\mathbf{y}_3\}.$$

By the methods of Ex. 3.7 and Ex. 3.8 we find

$$P = \begin{bmatrix} 1 & 0 & 1 & -2 \\ 0 & 1 & 4 & 1 \\ 0 & 0 & -3 & 0 \\ 0 & 0 & 1 & 1 \end{bmatrix} \text{ and } H^{-1} = \begin{bmatrix} 0 & 0 & -1 \\ 0 & -1 & 0 \\ 1 & 2 & 1 \end{bmatrix};$$

we then find

$$H^{-1}AP = \left[\begin{array}{cc|cc} 1 & 0 & 0 & 0 \\ 0 & 1 & 0 & 0 \\ \hline 0 & 0 & 0 & 0 \end{array}\right]$$

The proof of the following theorem depends on Theorem 4.3 and its corollary.

Theorem 4.6
The matrix A of order $n \times n$ is non-singular if and only if it has rank n.

Proof

(i) Assume A is non-singular and has rank r. Since A is of order $n \times n$, it represents a linear transformation

$$T : U \to U$$

where U is an n-dimensional vector space.

$\text{Im}(T) = T(U) = W_1$ (say) where W_1 is a subspace of U, and by the above assumption

$$r = r(A) = \text{Rank}(T) = \dim \text{Im}(T) = \dim W_1.$$

Since W_1 is a subspace of U

$$r \leq n. \tag{4.14}$$

Since A is a non-singular it has an inverse A^{-1} or order $n \times n$ which represents the linear transformation T^{-1} such that

$$T^{-1}T = 1 \quad \text{(the identity transformation)}$$

Also $T^{-1}(W_1) = W_2$ (say) where W_2 is a subspace of W_1 of dimension s (say), hence since $\dim W_2 \leq \dim W_1$, we have

$$s \leq r. \tag{4.15}$$

On the other hand

$$W_2 = T^{-1}(W_1) = T^{-1}(T(U)) = T^{-1}T(U) = U,$$

hence

$$s = \dim W_2 = \dim U = n.$$

Combining this result with (4.14) and (4.15), we can conclude that

$$r = n.$$

(ii) Assume $r(A) = n$. A is a matrix of a linear transformation

$$T : U \to V$$

where $\dim U = n$.
Since A has rank n,

$$n = \dim \operatorname{Im}(T),$$

that is, $\dim V = n$.
By Theorem 3.3, T is an isomorphism and so T^{-1} exists. It follows that A^{-1}, the matrix representing T^{-1}, exists, hence A is non-singular.

Theorem 4.7
If the matrix product AB exists, then

$$r(AB) \leq r(A)$$

Proof
Assume that A is of order $m \times p$, B is of order $p \times n$, so that AB is of order $m \times n$. We shall use the notation $L(AB)$ to denote the space spanned by the columns of the matrix (AB).

For whatever $\mathbf{y} \in L(AB)$, there exists $\mathbf{x} \in R^n$ such that

$$\mathbf{y} = AB\mathbf{x}. \tag{4.16}$$

Let $\mathbf{z} = B\mathbf{x}$, then (4.16) becomes

$$\mathbf{y} = A\mathbf{z} \tag{4.17}$$

that is, $\mathbf{y} \in L(A)$. Hence

$$L(AB) \subseteq L(A) \tag{4.18}$$

that is, $r(AB) \leq r(A)$.

Corollary
If B is non-singular, $r(AB) = r(A)$. If $\mathbf{y} \in L(A)$, $\mathbf{y} = A\mathbf{z}$ for some $\mathbf{z} \in R^p$. We can write

$$\mathbf{y} = ABB^{-1}\mathbf{z} = AB\mathbf{w} \text{ where } \mathbf{w} = B^{-1}\mathbf{z}.$$

So whenever $\mathbf{y} \in L(A)$, $\mathbf{y} \in L(AB)$.

That is,

$$L(A) \supseteq L(AB). \tag{4.19}$$

(4.18) and (4.19) together imply that

$$r(A) = r(AB)$$

In this section we have considered certain properties of the rank of matrices, but not the practical aspect – the computation of the rank. The problem can be solved by determining the number of linearly independent rows or columns of the matrix – this in turn lead us to solve the systems of simultaneous equations.

Another way of determining the rank of a matrix is by the use of so-called 'elementary operations'. We shall consider this technique in Chapter 8.

4.3 THE DETERMINANT OF A MATRIX

In this section we shall briefly develop some properties of determinants and define the basic manipulations, although it will be assumed that the reader is familiar with the more elementary aspects of determinant theory.

We associate a determinant with *square* matrix A, it is a single number, usually denoted by $|A|$ or Δ, representing, in some way to be discussed, the numerical value of the n^2 components of $A = [a_{ij}]$.

We shall define the determinant of order n in terms of determinants of order $(n-1)$. It will be necessary for us to define the determinant of the 1×1 order matrix $A = [a_{11}]$ as the value of this element,

that is,

$$|A| = a_{11}.$$

It will be simpler to define the value of a determinant in terms of its *minors* or *co-factors.*

Definition 4.5

(1) The **minor** denoted by M_{ij} of the a_{ij} element of the determinant of the nth order matrix A, is the $(n-1)^{\text{st}}$ order determinant of the matrix obtained by removing the ith row and the jth column from A.

(2) The **co-factor**, denoted by A_{ij}, of the element a_{ij} is the number

$$A_{ij} = (-1)^{i+j} M_{ij}.$$

The co-factor of the element a_{ij} is therefore the minor of that element, multiplied by $+1$ or -1 according to the position of the element as suggested by the following checkboard pattern:

$$\begin{bmatrix} + & - & + & \dots \\ - & + & - & \dots \\ + & - & + & \dots \\ \cdot & \cdot & \cdot & \end{bmatrix}.$$

Example 4.6

$$\text{If } |A| = \begin{vmatrix} a_{11} & a_{12} & a_{13} \\ a_{21} & a_{22} & a_{23} \\ a_{31} & a_{32} & a_{33} \end{vmatrix}$$

find M_{22}, M_{31}, A_{13} and A_{32}.

Solution

On deleting the second row and column from $|A|$

$$\begin{vmatrix} a_{11} & a_{12} & a_{13} \\ a_{21} & a_{22} & a_{23} \\ a_{31} & a_{32} & a_{33} \end{vmatrix} \text{ we are left with } \begin{vmatrix} a_{11} & a_{13} \\ a_{31} & a_{33} \end{vmatrix} = M_{22}.$$

Similarly

$$M_{31} = \begin{vmatrix} a_{12} & a_{13} \\ a_{22} & a_{23} \end{vmatrix}.$$

Obtaining the minors and multiplying by +1 or −1 according to the checkboard pattern, we find

$$A_{13} = (+1) \begin{vmatrix} a_{21} & a_{22} \\ a_{31} & a_{32} \end{vmatrix} \text{ and } A_{32} = (-1) \begin{vmatrix} a_{11} & a_{13} \\ a_{21} & a_{23} \end{vmatrix}.$$

Definition 4.6
The determinant of order n of the matrix A has the value

$$|A| = a_{11}A_{11} + a_{12}A_{12} + \ldots + a_{1n}A_{1n}$$

where $a_{11}, a_{12}, \ldots a_{1n}$ are the elements of the first row of A and $A_{11}, A_{12}, \ldots A_{1n}$ are the corresponding cofactors.

For example

$$|A| = \begin{vmatrix} a_{11} & a_{12} \\ a_{21} & a_{22} \end{vmatrix} = a_{11}A_{11} + a_{12}A_{12} = a_{11}a_{22} + a_{12}(-1)a_{21}$$
$$= a_{11}a_{22} - a_{12}a_{21}$$

and

$$|A| = \begin{vmatrix} a_{11} & a_{12} & a_{13} \\ a_{21} & a_{22} & a_{23} \\ a_{31} & a_{32} & a_{33} \end{vmatrix} = a_{11}A_{11} + a_{12}A_{12} + a_{13}A_{13} \qquad (4.20)$$

$$= a_{11}\begin{vmatrix} a_{22} & a_{23} \\ a_{32} & a_{33} \end{vmatrix} - a_{12}\begin{vmatrix} a_{21} & a_{23} \\ a_{31} & a_{33} \end{vmatrix} + a_{13}\begin{vmatrix} a_{21} & a_{22} \\ a_{31} & a_{32} \end{vmatrix}.$$

It is not very difficult to prove that the determinant is the sum of the products of the elements of *any* row, by their corresponding co-factors, thus (4.20) has the following equivalent expansions:

$$|A| = a_{11}A_{11} + a_{12}A_{12} + a_{13}A_{13}$$
$$= a_{21}A_{21} + a_{22}A_{22} + a_{23}A_{23}$$
$$= a_{31}A_{31} + a_{32}A_{32} + a_{33}A_{33}.$$

Example 4.7
Find the value of the determinant $\begin{vmatrix} 1 & 2 & -2 \\ -1 & 1 & 3 \\ 2 & -1 & 2 \end{vmatrix}.$

Solution
Expanding by the first row

$$\Delta = 1\begin{vmatrix} 1 & 3 \\ -1 & 2 \end{vmatrix} - 2\begin{vmatrix} -1 & 3 \\ 2 & 2 \end{vmatrix} - 2\begin{vmatrix} -1 & 1 \\ 2 & -1 \end{vmatrix}$$

$$= (2 + 3) - 2(-2 - 6) - 2(1 - 2) = 23.$$

We can expand by either of the remaining rows, for example by the third:

$$\Delta = 2\begin{vmatrix} 2 & -2 \\ 1 & 3 \end{vmatrix} + \begin{vmatrix} 1 & -2 \\ -1 & 3 \end{vmatrix} + 2\begin{vmatrix} 1 & 2 \\ -1 & 1 \end{vmatrix}$$

$$= 2(6 + 2) + (3 - 2) + 2(1 + 2) = 23.$$

The following further properties of determinants are stated without proof. Some of them are verified in examples

I. If two rows (or columns) of a matrix A are interchanged, to become the matrix B, then

$$|B| = -|A|.$$

II. If A' is the transpose of A, then

$$|A'| = |A|.$$

III. If the elements of one row (or column) of A are multiplied by a constant c, the value of the determinant of the new matrix is

$$c|A|.$$

IV. If any row (or column) of A is a multiple of any other row (or column) of A, then

$$|A| = 0.$$

Example 4.8
Evaluate

(1) $\begin{vmatrix} 2 & -4 \\ -1 & 2 \end{vmatrix}$ (2) $\begin{vmatrix} 2 & 1 & -2 \\ 1 & -1 & 3 \\ -1 & 2 & 2 \end{vmatrix}$

(3) $\begin{vmatrix} 2 & 1 & -2 \\ 3 & -3 & 9 \\ -1 & 2 & 2 \end{vmatrix}$, and (4) write as a product of linear factors

$$\begin{vmatrix} 1 & 1 & 1 \\ x & y & z \\ x^2 & y^2 & z^2 \end{vmatrix}.$$

Solution

(1) Notice that the second column is a multiple (–2) of the first column, hence by (IV) $\Delta = 0$. Indeed $\Delta = 2.2 - (-4)(-1) = 0$.

(2) This determinant is the same as in Ex. 4.4, except that columns (1) and (2) have been interchanged. By *I*, $\Delta = -23$. Indeed

$$\Delta = 2(-2 - 6) - 1(2 + 3) - 2(2 - 1) = -23.$$

(3) This determinant is recognised to be the same as in (2) except that the second row has been multiplied by 3.
By (III), $\Delta = 3(-23) = -69$. Indeed

$$\Delta = 2(-6 - 18) - 1(6 + 9) - 2(6 - 3) = -69.$$

(4) If we allow $x = y$, the determinant will have two equal columns and then $\Delta = 0$ by (IV). Hence (by the Remainder Theorem) $(x - y)$ is a factor of Δ. Similarly $(y - z)$ and $(z - x)$ are factors. Hence Δ must be of the form

$$\Delta = c(x - y)(y - z)(z - x), c \text{ being a constant.}$$

Comparing the coefficients of yz^2 (say) along the leading diagonal of Δ and in the above expression, we find that $c = 1$.

4.4 OPERATIONS WITH DETERMINANTS

In this section we shall be concerned with various operations with determinants, but will not attempt formal proofs of the various results, although we shall verify them.

Theorem 4.8
If a matrix B is formed from a matrix A by adding to any row (column) of A, k times the corresponding elements in any other row (column), then

$$|B| = |A|.$$

For example

$$\text{if } A = \begin{bmatrix} a & b \\ c & d \end{bmatrix} \text{ and } B = \begin{bmatrix} a & b \\ c+ka & d+kb \end{bmatrix}$$

then

$$\begin{aligned} |B| &= a(d + kb) - b(c + ka) \\ &= (ad - bc) + k(ab - ab) \\ &= \begin{vmatrix} a & b \\ c & d \end{vmatrix} = |A|. \end{aligned}$$

We can generalise this result to the following type of expansion

$$\begin{vmatrix} a_1 + x_1 & b_1 & c_1 \\ a_2 + x_2 & b_2 & c_2 \\ a_3 + x_3 & b_3 & c_3 \end{vmatrix} = \begin{vmatrix} a_1 & b_1 & c_1 \\ a_2 & b_2 & c_2 \\ a_3 & b_3 & c_3 \end{vmatrix} + \begin{vmatrix} x_1 & b_1 & c_1 \\ x_2 & b_2 & c_2 \\ x_3 & b_3 & c_3 \end{vmatrix}.$$

The following theorem specifies the rule for determinant multiplication.

Theorem 4.9
Given two matrices A,B of order $n \times n$.

$$|A| \,.\, |B| = |A.B| \tag{4.19}$$

As mentioned above, we shall not attempt a rigorous proof.

If $A = \begin{bmatrix} a & b \\ c & d \end{bmatrix}$ and $B = \begin{bmatrix} x & y \\ z & w \end{bmatrix}$, then

$$AB = \begin{bmatrix} ax + bz & ay + bw \\ cx + cy & cy + dw \end{bmatrix}.$$

By Theorem 4.8 can write $|AB|$ as

$$|AB| = \begin{vmatrix} ax & ay \\ cx & cy \end{vmatrix} + \begin{vmatrix} ax & bw \\ cx & dw \end{vmatrix} + \begin{vmatrix} bz & ay \\ dz & cy \end{vmatrix} + \begin{vmatrix} bz & bw \\ dz & dw \end{vmatrix}.$$

Since $\begin{vmatrix} ax & ay \\ cx & cy \end{vmatrix} = ac \begin{vmatrix} x & y \\ x & y \end{vmatrix} = 0$

and $\begin{vmatrix} bz & bw \\ dz & dw \end{vmatrix} = bd \begin{vmatrix} z & w \\ z & w \end{vmatrix} = 0$, it follows that

$$|AB| = xw \begin{vmatrix} a & b \\ c & d \end{vmatrix} + yz \begin{vmatrix} b & a \\ d & c \end{vmatrix} = xw \begin{vmatrix} a & b \\ c & d \end{vmatrix} - yz \begin{vmatrix} a & b \\ c & d \end{vmatrix}$$

$$= \begin{vmatrix} a & b \\ c & d \end{vmatrix} (xw - yz) = \begin{vmatrix} a & b \\ c & d \end{vmatrix} \begin{vmatrix} x & y \\ z & w \end{vmatrix} = |A|.|B|.$$

In the example below we show that when the columns (rows) of a matrix are linearly dependent (that is, the rank of the matrix is smaller than its order and so the matrix is *singular*), the determinant of the matrix is 0.

Example 4.9
The matrix of A of order n has columns $\mathbf{a}_1, \mathbf{a}_2, \ldots \mathbf{a}_n$. If these vectors are linearly dependent, show that $|A| = 0$.

Solution

Since $\mathbf{a}_1, \mathbf{a}_2, \ldots \mathbf{a}_n$ are linearly dependent, $\mathbf{a}_i$ (say) is a linear combination (see Theorem 2.3) of the preceding vectors; that is,

$$\mathbf{a}_i = \alpha_1\mathbf{a}_1 + \alpha_2\mathbf{a}_2 + \ldots + \alpha_{i-1}\mathbf{a}_{i-1},$$

hence

$$|A| = |\mathbf{a}_1, \mathbf{a}_2, \ldots \mathbf{a}_{i-1}, \alpha_1\mathbf{a}_1 + \ldots + \alpha_{i-1}\mathbf{a}_{i-1}, \mathbf{a}_{i+2}, \ldots \mathbf{a}_n|$$

$$= \alpha_1|\mathbf{a}_1, \mathbf{a}_2, \ldots \mathbf{a}_{i-1}, \mathbf{a}_1, \ldots \mathbf{a}_n|$$

$$+ \alpha_2|\mathbf{a}_1, \mathbf{a}_2, \ldots \mathbf{a}_{i-1}, \mathbf{a}_2, \ldots \mathbf{a}_n|$$

$$\ldots$$

$$+ \alpha_{i-1}|\mathbf{a}_1, \mathbf{a}_2, \ldots \mathbf{a}_{i-1}, \mathbf{a}_{i-1}, \ldots \mathbf{a}_n|. \qquad \text{(by theorem 4.8)}$$

But each of these determinants is 0 by property (IV), hence $|A| = 0$. The converse of the above property, that is, that if $|A| = 0$, then $\mathbf{a}_1, \mathbf{a}_2, \ldots \mathbf{a}_n$ are linearly dependent, can also be proved.

4.5 CRAMER'S RULE

Solutions of simultaneous equations can be evaluated in terms of determinants by Cramer's rule. Before discussing it we need the following result which is summarised as a theorem.

Theorem 4.10

Given the matrix $A = [a_{ij}]$ of order $n \times n$ and the co-factors A_{ij} of the elements a_{ij}, then

$$a_{ij}A_{is} + a_{2j}A_{2s} + \ldots + a_{nj}A_{ns} = \begin{cases} |A| & \text{if } j = s \\ 0 & \text{if } j \neq s. \end{cases}$$

Discussion

We shall not attempt a rigorous proof. The fact that

$$\sum_{i=1}^{n} a_{ij}A_{ij} = |A| \tag{4.21}$$

has been verified in Sec. 4.3. To understand the result that

$$\sum_{i=1}^{n} a_{ij}A_{is} = 0 \quad (j \neq s) \tag{4.22}$$

we must think of this sum as the product of the elements in the jth column of the matrix A multiplied by the co-factors of the elements of the sth column of A.

This is just the situation we find when the matrix A has equal elements in the jth and sth columns, that is, if A has the form:

$$A = \begin{bmatrix} a_{11} & \dots & a_{1j} & \dots & a_{1j} & \dots & a_{1n} \\ a_{21} & \dots & a_{2j} & \dots & a_{2j} & \dots & a_{2n} \\ \vdots & & & & & & \\ a_{n1} & \dots & a_{nj} & \dots & a_{nj} & \dots & a_{nn} \end{bmatrix}.$$
$$\qquad \uparrow \qquad\qquad \uparrow$$
$$\qquad j^{\text{th}} \qquad\quad s^{\text{th}}$$

But we know (property (IV)) that the determinant of a matrix having two equal columns (rows) is 0. The result follows.

Example 4.10
Verify that

$$a_{11}A_{12} + a_{21}A_{22} + a_{31}A_{32} = 0$$

for the determinant of Ex. 4.7.

Solution

$$A_{12} = \begin{vmatrix} -1 & 3 \\ 2 & 2 \end{vmatrix} = -8 \quad A_{22} = \begin{vmatrix} 1 & -2 \\ 2 & 2 \end{vmatrix} = 6 \quad A_{32} = \begin{vmatrix} 1 & -2 \\ -1 & 3 \end{vmatrix} = 1,$$

hence

$$\sum_{i=1}^{3} a_{i1}A_{i2} = 1 \times (-8) - 1 \times 6 + 2 \times 1 = 0.$$

Now we can discuss Cramer's rule, which for simplicity we do for the case of 3 simultaneous equations in 3 unknowns. The case of n equations in n unknowns is a simple generalization.

Let

$$\begin{aligned} a_{11}x_1 + a_{12}x_2 + a_{13}x_3 &= b_1 \\ a_{21}x_1 + a_{22}x_2 + a_{23}x_3 &= b_2 \\ a_{31}x_1 + a_{32}x_2 + a_{33}x_3 &= b_3 \end{aligned} \tag{4.23}$$

be represented by the matrix equation

$$A\mathbf{x} = \mathbf{b} \tag{4.24}$$

for which it is assumed that $\Delta = |A| \neq 0$.

We multiply the first equation of (4.23) by A_{11}, the second by A_{21}, and the third by A_{31} and add. We obtain

$$(a_{11}A_{11} + a_{21}A_{21} + a_{31}A_{31})x_1 = b_1A_{11} + b_2A_{21} + b_3A_{31} \tag{4.25}$$

(the other two terms on the left-hand side are 0 by (4.22)).

But $b_1A_{11} + b_2A_{21} + b_3A_{31} = \begin{vmatrix} b_1 & a_{12} & a_{13} \\ b_2 & a_{22} & a_{23} \\ b_3 & a_{32} & a_{33} \end{vmatrix}$,

and since we have assumed that $\Delta \neq 0$, we obtain from (4.25)

$$x_1 = \frac{\begin{vmatrix} b_1 & a_{12} & a_{13} \\ b_2 & a_{22} & a_{23} \\ b_3 & a_{32} & a_{33} \end{vmatrix}}{\begin{vmatrix} a_{11} & a_{12} & a_{13} \\ a_{21} & a_{22} & a_{23} \\ a_{31} & a_{32} & a_{33} \end{vmatrix}} = \frac{\Delta_1}{\Delta}$$

where Δ_1 is the determinant of the matrix obtained by replacing the first column of A by $\mathbf{b}$. Similarly we obtain

$$x_2 = \frac{\Delta_2}{\Delta} \quad \text{and} \quad x_3 = \frac{\Delta_3}{\Delta}.$$

In the general case, Cramer's rule is written as

$$x_i = \frac{\Delta_i}{\Delta} \quad (i = 1,2, \dots n) \tag{4.26}$$

where Δ (assumed $\neq 0$) is the determinant of the matrix A, and Δ_i is the determinant of the matrix obtained by replacing the ith column of A by $\mathbf{b}$.

Example 4.11
Use Cramer's rule to solve the following system of equations:

$$\begin{aligned} x_1 + 2x_2 - 2x_3 &= -7 \\ -x_1 + x_2 + 3x_3 &= 3 \\ 2x_1 - x_2 + 2x_3 &= 8. \end{aligned}$$

Solution

$$\Delta = \begin{vmatrix} 1 & 2 & -2 \\ -1 & 1 & 3 \\ 2 & -1 & 2 \end{vmatrix} = 23 \quad \text{(see Ex. 4.7)}$$

$$\Delta_1 = \begin{vmatrix} -7 & 2 & -2 \\ 3 & 1 & 3 \\ 8 & -1 & 2 \end{vmatrix} = 23 \qquad \Delta_2 = \begin{vmatrix} 1 & -7 & -2 \\ -1 & 3 & 3 \\ 2 & 8 & 2 \end{vmatrix} = -46$$

$$\text{and } \Delta_3 = \begin{vmatrix} 1 & 2 & -7 \\ -1 & 1 & 3 \\ 2 & -1 & 8 \end{vmatrix} = 46$$

hence $x_1 = \frac{23}{23} = 1$, $x_2 = -\frac{46}{23} = -2$, and $x_3 = \frac{46}{23} = 2$.

There are of course other methods used for solving systems of equations. One direct method is to find the inverse matrix A^{-1} so that (4.24) can be written as

$$\mathbf{x} = A^{-1}\mathbf{b}.$$

We shall discuss a number of methods for inverting matrices in Chapter 8. But one direct method uses the *adjoint* matrix of A.

Definition 4.7
The transpose of the matrix of co-factors of the elements of the square matrix $A = [a_{ij}]$ is known as the **adjoint** of A, denoted by adj A, that is,

$$\text{if } A = \begin{bmatrix} a_{11} & a_{12} & \dots & a_{1n} \\ a_{21} & a_{22} & \dots & a_{2n} \\ \vdots & & & \\ a_{n1} & a_{n2} & \dots & a_{nn} \end{bmatrix} \quad \text{then adj } A = \begin{bmatrix} A_{11} & A_{21} & \dots & A_{n1} \\ A_{12} & A_{22} & \dots & A_{n2} \\ \vdots & & & \\ A_{1n} & A_{2n} & \dots & A_{nn} \end{bmatrix}.$$

We summarise the method for inverting a non-singular matrix in the following theorem:

Theorem 4.11
If A is a non-singular square matrix, then

$$A^{-1} = \frac{1}{|A|}\text{adj } A. \tag{4.27}$$

Proof
To find the product of A and adj A we carry out the normal matrix multiplication. Thus starting with the first row of A by the first column of adj A, we obtain

$$a_{11}A_{11} + a_{12}A_{12} + \dots + a_{1n}A_{1n} = |A|.$$

The product of the first row of A with each of the remaining columns of adj A is a sum of (4.21) form, that is

$$\sum_{i=1}^{n} a_{1i}A_{ri} = 0 \quad (r = 2,3,\dots n).$$

We obtain similar results for the products of the remaining rows of A with the columns of adj A, so that

$$A \operatorname{adj} A = \begin{bmatrix} |A| & 0 & \dots & 0 \\ 0 & |A| & \dots & 0 \\ \vdots & & & \\ 0 & 0 & \dots & |A| \end{bmatrix} = |A| \begin{bmatrix} 1 & 0 & \dots & 0 \\ 0 & 1 & \dots & 0 \\ \vdots & & & \\ 0 & 0 & \dots & 1 \end{bmatrix}.$$

But $|A| \neq 0$, hence

$$\frac{A \operatorname{adj} A}{|A|} = \mathrm{I},$$

and (4.27) follows.

Example 4.12
Find A^{-1} and hence solve the system of equations of Ex. 4.11.

Solution

$$\text{Since } A = \begin{bmatrix} 1 & 2 & -2 \\ -1 & 1 & 3 \\ 2 & -1 & 2 \end{bmatrix} \text{ we find } \operatorname{adj} A = \begin{bmatrix} 5 & -2 & 8 \\ 8 & 6 & -1 \\ -1 & 5 & 3 \end{bmatrix}$$

$$\mathbf{x} = \begin{bmatrix} x_1 \\ x_2 \\ x_3 \end{bmatrix} = \tfrac{1}{23} \begin{bmatrix} 5 & -2 & 8 \\ 8 & 6 & -1 \\ -1 & 5 & 3 \end{bmatrix} \begin{bmatrix} -7 \\ 3 \\ 8 \end{bmatrix} = \begin{bmatrix} 1 \\ -2 \\ 2 \end{bmatrix}.$$

Note
In practice it is generally unnecessary to invert a non-singular matrix A to obtain the solution of $A\mathbf{x} = \mathbf{b}$. There are various methods both direct and indirect for solving such systems of equations. References 7, 10, and 18 among others give details of such procedures.

PROBLEMS FOR CHAPTER 4

1) $T : R^3 \to R^2$ is a linear transformation defined by

$$T([x,y,z]) = [x-2y,\ 2x-4y]\,.$$

Find
(a) the basis of ker T and the nullity of T,
(b) the basis of Im(T) and the rank of T, and
(c) verify the Dimension Theorem.

2) Determine (i) the basis and (ii) the dimensions for ker T and Im(T) when T is represented by

(i)
$$A = \begin{bmatrix} -1 & 2 & 0 \\ 1 & -1 & 0 \end{bmatrix};$$

(ii)
$$A = \begin{bmatrix} -1 & -1 & -3 & -3 \\ 0 & 1 & 1 & 2 \\ 1 & 0 & 2 & 1 \end{bmatrix}.$$

3) Find the rank of the matrix

$$\begin{bmatrix} 3 & 2 & 5-x \\ 1 & -1 & -1+x \\ 2+x & 3 & 4 \end{bmatrix}.$$

for $x \in R$.

4) Given

$$A = \begin{bmatrix} 1 & -1 & 1 & 0 \\ 2 & 1 & 5 & 3 \\ -1 & 3 & 1 & 2 \\ 0 & 1 & 1 & 1 \end{bmatrix},$$

find the matrices H and P such that

$H^{-1}AP$ is in the form

$$\left[\begin{array}{c:c} I & 0 \\ \hdashline 0 & 0 \end{array}\right].$$

5) A system whose state equations (see Sec. 1.5) are

$$\dot{\mathbf{x}} = A\mathbf{x} + Bu$$

$$y = C\mathbf{x}$$

where $\mathbf{x}$ is of order n, A is an $(n \times n)$ matrix, B is an $(n \times 1)$ matrix C is a $(1 \times n)$ matrix, is said to be **completely controllable** if the partitioned matrix of order $(n \times n)$

$$[B \mid AB \mid \ldots \mid A^{n-1}B]$$

has rank n.

The system is said to be **completely observable** if the partitioned matrix of order $(n \times n)$

$$\left[\begin{array}{c} C \\ \hdashline CA \\ \hdashline \vdots \\ \hdashline CA^{n-1} \end{array}\right]$$

has rank n.

Given the system

$$\begin{bmatrix} \dot{x}_1 \\ \dot{x}_2 \\ \dot{x}_3 \end{bmatrix} = \begin{bmatrix} 0 & 1 & 0 \\ 1 & 0 & -1 \\ -1 & -2 & -3 \end{bmatrix} \begin{bmatrix} x_1 \\ x_2 \\ x_3 \end{bmatrix} + \begin{bmatrix} 0 \\ 1 \\ 0 \end{bmatrix} [u]$$

$$y = [1\ 0\ 1]\begin{bmatrix} x_1 \\ x_2 \\ x_3 \end{bmatrix},$$

determine whether it is completely controllable and completely observable.

6) Use Cramer's Rule to solve the system of equations

$$\begin{aligned} 3x_1 + 2x_2 - x_3 &= -1 \\ x_1 - 2x_2 + 2x_3 &= 7 \\ 2x_1 + x_2 + x_3 &= 3 \end{aligned}\quad .$$

7) Find the adjoint of A where A is the matrix of coefficients in Ex. 4.6. Hence evaluate the inverse of A.

8) In Sec. 2.1, the vector product of two vectors was defined. We can write this product in the form of a determinant. If **i**,**j**,**k** are the unit vectors (1, 0, 0), (0, 1, 0), (0, 0,1) respectively, we can write the product as

$$\mathbf{x} \times \mathbf{y} = \begin{vmatrix} \mathbf{i} & \mathbf{j} & \mathbf{k} \\ x_1 & x_2 & x_3 \\ y_1 & y_2 & y_3 \end{vmatrix} = \mathbf{i}(x_2y_3 - y_2x_3) + \mathbf{j}(x_3y_1 - x_1y_3) + \mathbf{k}(x_1y_2 - x_2y_1).$$

Let $\mathbf{x} = (1,2,-1)$, $\mathbf{y} = (2,1,3)$ and $\mathbf{z} = (2,1,2)$.
Determine

(i) $\mathbf{x} \times \mathbf{y}$, and

(ii) $(\mathbf{x} \times \mathbf{y}) \times \mathbf{z}$.

9) Show that the values of the following determinants are independent of a

(i)
$$\begin{vmatrix} 1 & a+3 & (a+2)(a+3) \\ 1 & a+4 & (a+3)(a+4) \\ 1 & a+5 & (a+4)(a+5) \end{vmatrix}$$

(ii)

$$\begin{vmatrix} 1 & a+2 & (a+2)^2 \\ 1 & a+3 & (a+3)^2 \\ 1 & a+4 & (a+4)^2 \end{vmatrix}$$

10) Show that

$$|I + \mathbf{x}\mathbf{y}'| = 1 + \mathbf{x} \cdot \mathbf{y}$$

in the case when I is the unit matrix of order 2×2 and **x** and **y** are vectors of order 2. (The result is true in the general case when I, **x** and **y** are of order n).

CHAPTER 5

Linear Equations

In the last chapter we discussed a method, Cramer's rule, for solving a set of simultaneous equations, when it was known (by the assumption that $\Delta \neq 0$) that such a solution does exist.

In this chapter we shall consider more general problems – whether a solution exists and if it does, is the solution unique?

5.1 SYSTEMS OF HOMOGENEOUS EQUATIONS

A system of equations is **homogeneous** when the constant terms, that is, the right-hand sides, are zero. Such a system consisting of m equations in n unknowns: $x_1, x_2, \ldots x_n$ is of the form

$$\begin{aligned} a_{11}x_1 + a_{12}x_2 + \ldots + a_{1n}x_n &= 0 \\ a_{21}x_1 + a_{22}x_2 + \ldots + a_{2n}x_n &= 0 \\ &\vdots \\ a_{m1}x_1 + a_{m2}x_2 + \ldots + a_{mn}x_n &= 0. \end{aligned} \tag{5.1}$$

In matrix form (5.1) is written as

$$A\mathbf{x} = \mathbf{0} \tag{5.2}$$

where $A = [a_{ij}]$ is of order $m \times n$, and $\mathbf{x} = [x_1, x_2, \ldots x_n]'$. The right-hand side of (5.2) is the zero vector of order m.

Definition 5.1

A system of equations is said to be **consistent** if it has at least one solution, otherwise the system is said to be **inconsistent**.

Definition 5.2
If it exists, a non-zero solution $\mathbf{x}$ (say) to (5.2) is called a **non-trivial** solution.

Notice that $\mathbf{x} = [0,0,\ldots 0]'$ is always a solution to (5.2); it is known as the **trivial** solution.

Also, if $\mathbf{x}$ satisfies (5.2) so does $\lambda\mathbf{x}$ where λ is any scalar, since

$$A(\lambda\mathbf{x}) = \lambda(A\mathbf{x}) = \lambda\mathbf{O} = \mathbf{O}.$$

This implies that if a non-trivial solution does exist, it is not unique; indeed, there are an 'infinity of solutions'.

The above is an interesting observation; it means that a system such as (5.2) can never be inconsistent; it always has at least one solution. The question which we shall now attempt to answer is under what conditions does (5.2) have non-trivial solutions, and how many of these solutions are linearly independent.

We consider A in (5.2) as the matrix representing a linear transformation $T : U \to V$, relative to the standard bases, where U and V have dimensions n and m respectively. Any solution of (5.2) is in the kernel of T (see Def. 4.1), and by the dimension theorem

$$\begin{aligned} \dim \ker T &= n - \dim \operatorname{Im}(T) \\ &= n - r \end{aligned} \tag{5.3}$$

where r is the rank of A.

Eq. (5.3) answers our question, which we summarise in the following theorem:

Theorem 5.1
The system of equations $A\mathbf{x} = \mathbf{O}$, where A is a matrix of order $m \times n$ and of rank r, has

(i) a unique trivial solution $\mathbf{x} = \mathbf{O}$, if

$$r = n, \text{ or}$$

(ii) $(n-r)$ linearly independent non-trivial solutions if

$$r < n.$$

Proof

(i) If $r = n$, $\dim \ker T = n - n = 0$ (see (5.3)), hence $\ker T$ consists of the one element $\{\mathbf{O}\}$.

(ii) If $n > r$, $\ker T$ has dimension $(n-r)$ and so must be spanned by $(n-r)$ linearly independent vectors – each one of which satisfies (5.2).

We need to interpret Theorem 5.1 in the three possible cases.

Case 1: The number of equations is greater than the number of unknowns

$$(m > n).$$

Case 2: The number of equations is equal to the number of unknowns

$$(m = n).$$

Case 3: The number of equations is smaller than the number of unknowns

$$(m < n).$$

Case 1: $m > n$

A very simple example of this situation is the following:

$$2x_1 - x_2 = 0, x_1 + 2x_2 = 0, 3x_1 + x_2 = 0.$$

One of these equations is **redundant**, which means that all the information contained in it can be obtained from the other two. Indeed the third equation is the sum of the other two. All the redundant equations can be discarded, leaving us with a system of homogeneous equations for which $m = n$ or $m < n$.

Case 2: $m = n$

There are two cases to consider

(a) A is non-singular.
By Theorem 4.6, the rank of A is n, hence $\mathbf{x} = \mathbf{O}$ is the unique solution (Theorem 5.1).

(b) A is singular, $r(A) = r < n$.
By Theorem 5.1, the system of equations (5.2) has $(n-r)$ linearly independent (non-trivial) solutions.

Case 3: $m < n$

Let $r(A) = r$.

By Theorem 4.3, $r \leq m$.

But since $m < n$, $n - r > 0$, and in this case there are $(n-r)$ linearly independent (non-trivial) solutions.

It is clear that the most interesting cases occur when A is singular and when $m < n$.

The next problem to consider is how to solve the system of equations in these cases.

If A is a matrix of order $m \times n$, of rank r, and $\mathbf{x} = [x_1, x_2, \ldots x_n]'$ is any vector of order n,

$$A\mathbf{x} = \mathbf{O}$$

can be written (see (4.4)) as

$$x_1\mathbf{a}_1 + x_2\mathbf{a}_2 + \ldots + x_n\mathbf{a}_n = \mathbf{O} \tag{5.4}$$

where $\mathbf{a}_1,\mathbf{a}_2,\ldots\mathbf{a}_n$ are the column vectors of A. There are (see Def. 4.3) r linearly independent vectors among $\mathbf{a}_1,\mathbf{a}_2,\ldots\mathbf{a}_n$. We shall assume that the r vectors $\mathbf{a}_1,\mathbf{a}_2,\ldots\mathbf{a}_r$ are linearly independent, the remaining vectors $\mathbf{a}_{r+1}, \mathbf{a}_{r+2}, \ldots \mathbf{a}_n$ being dependent on the first r vectors. This assumption leads to a simplification in the notation which would need to be used otherwise, and this outweighs the disadvantages of the slight loss in generality.

For an **arbitrary** choice of $x_{r+1},x_{r+2},\ldots x_n$ the vector

$$(x_{r+1}\mathbf{a}_{r+1} + x_{r+2}\mathbf{a}_{r+2} + \ldots + x_n\mathbf{a}_n)$$

can be expressed as a **unique** linear combination of $\mathbf{a}_1,\mathbf{a}_2,\ldots\mathbf{a}_r$, that is, there exist uniquely determined numbers $\beta_1,\beta_2,\ldots\beta_r$ such that

$$x_{r+1}\mathbf{a}_{r+1} + \ldots + x_n\mathbf{a}_n = \beta_1\mathbf{a}_1 + \beta_2\mathbf{a}_2 + \ldots + \beta_r\mathbf{a}_r. \tag{5.5}$$

Substituting (5.5) into (5.4) we obtain

$$x_1\mathbf{a}_1 + \ldots + x_r\mathbf{a}_r + \beta_1\mathbf{a}_1 + \ldots + \beta_r\mathbf{a}_r = \mathbf{O},$$

that is,

$$(x_1 + \beta_1)\mathbf{a}_1 + \ldots + (x_r + \beta_r)\mathbf{a}_r = \mathbf{O}.$$

Since $\mathbf{a}_1,\ldots\mathbf{a}_r$ are linearly independent

$$x_1 = -\beta_1 \ldots x_r = -\beta_r.$$

We have shown that the solution to (5.4) consists of an *arbitrary* choice of the $(n-r)$ numbers $x_{r+1},x_{r+2},\ldots x_n$, the remaining r values $x_1,\ldots x_r$ being uniquely determined.

Notice that the above discussion is applicable to the case when A is a singular matrix of order $n \times n$, hence we have covered both of the 'interesting' cases.

Example 5.1
Solve the system of equations

$$\begin{aligned} 2x + y - z - w &= 0 \\ -x + 2y + 3z + 3w &= 0 \\ x + 8y + 7z + 7w &= 0. \end{aligned}$$

Solution

$$A = \begin{bmatrix} 2 & 1 & -1 & -1 \\ -1 & 2 & 3 & 3 \\ 1 & 8 & 7 & 7 \end{bmatrix}.$$

If $\mathbf{a}_1, \mathbf{a}_2, \mathbf{a}_3, \mathbf{a}_4$ are the columns of A, we note that

$$\mathbf{a}_3 = \mathbf{a}_2 - \mathbf{a}_1 = \mathbf{a}_4, \text{ so } r(A) = 2.$$

Since $n - r = 4 - 2$, there should be 2 non-trivial (linearly independent) solutions. The other implication is that one of the equations is redundant and could be discarded. For example, it can be verified that

(third equation) = 2(first equation) + 3(second equation).

Indeed the third equation (which we do not discard) does not contribute to the calculations which follow.

The solutions are vectors $[x, y, z, w]'$ such that

$$x\mathbf{a}_1 + y\mathbf{a}_2 + z\mathbf{a}_3 + w\mathbf{a}_4 = \mathbf{0}.$$

If we choose $z = 1$ and $w = 2$ (say), then

$$x\begin{bmatrix} 2 \\ -1 \\ 1 \end{bmatrix} + y\begin{bmatrix} 1 \\ 2 \\ 8 \end{bmatrix} + \begin{bmatrix} -1 \\ 3 \\ 7 \end{bmatrix} + 2\begin{bmatrix} -1 \\ 3 \\ 7 \end{bmatrix} = \begin{bmatrix} 0 \\ 0 \\ 0 \end{bmatrix},$$

that is,

$$x\begin{bmatrix}2\\-1\\1\end{bmatrix}+y\begin{bmatrix}1\\2\\8\end{bmatrix}=\begin{bmatrix}3\\-9\\-21\end{bmatrix}$$

On solving these equations, we find $x = 3$ and $y = -3$. Hence $[3, -3, 1, 2]'$ is one solution.

Making another arbitrary choice, say $z = 1, w = 0$, we find that

$$x\begin{bmatrix}2\\-1\\1\end{bmatrix}+y\begin{bmatrix}1\\2\\8\end{bmatrix}=\begin{bmatrix}1\\-3\\-7\end{bmatrix},$$

that is, $y = -1$ and $x = 1$.
Hence $[1, -1, 1, 0]'$ is another solution.

These two solutions are linearly independent, and any other solution must be a linear combination of these two.

As an example, if we choose $z = 0$ and $w = 5$ we find the solution $[5, -5, 0, 5]'$, and this can be written as the combination

$$\frac{5}{2}[3, -3, 1, 2]' - \frac{5}{2}[1, -1, 1, 0]'.$$

Definition 5.3
If $S = \{\mathbf{x}_1, \mathbf{x}_2, \ldots \mathbf{x}_{n-r}\}$ is a set of $(n-r)$ linearly independent solutions of the system $A\mathbf{x} = \mathbf{0}$, and if any solution of the system is a linear combination of the vectors in S, then S is called a **fundamental set of solutions.**

5.2 SYSTEMS OF NON-HOMOGENEOUS EQUATIONS

The system of m equations in n unknowns

$$\begin{aligned}
a_{11}x_1 + a_{12}x_2 + \ldots + a_{1n}x_n &= b_1\\
a_{21}x_1 + a_{22}x_2 + \ldots + a_{2n}x_n &= b_2\\
\vdots \qquad\qquad & \quad \vdots\\
a_{m1}x_1 + a_{m2}x_2 + \ldots + a_{mn}x_n &= b_m
\end{aligned}$$

is a **non-homogeneous** system if at least one of the b_i's (the right-hand side) is non-zero.

In matrix form the system is written as

$$A\mathbf{x} = \mathbf{b}. \tag{5.6}$$

$A = [a_{ij}]$ is a matrix of order $m \times n$
$\mathbf{x} = [x_1, x_2, \ldots x_n]'$ and $\mathbf{b} = [b_1, b_2, \ldots b_m]'$.

As in the case of homogeneous equations we shall first investigate conditions for which the equations are consistent, and then discuss the uniqueness problem.

We assume that the rank of A is r. Using the notation of the last section, we can write

$$A\mathbf{x} = x_1\mathbf{a}_1 + x_2\mathbf{a}_2 + \ldots + x_n\mathbf{a}_n. \tag{5.7}$$

Since the columns of A span the image space of T (the linear transformation whose matrix representation is A) a solution to (5.6) will exist only if $\mathbf{b}$ is in that space.

Conversely, if $\mathbf{b} \in \mathrm{Im}\,(T)$, that is, the space spanned by $\mathbf{a}_1, \mathbf{a}_2, \ldots \mathbf{a}_n$ then there exist scalars $x_1, x_2, \ldots x_n$ such that

$$x_1\mathbf{a}_1 + x_2\mathbf{a}_2 + \ldots + x_n\mathbf{a}_n = \mathbf{b}.$$

We summarise this result as a theorem.

Theorem 5.2
The system of equations (5.6) for which $r(A) = r$ will have a solution if and only if $\mathbf{b}$ is an element of the r-dimensional space spanned by the column vectors of the matrix A.

The fact that a solution exists only if $\mathbf{b}$ is an element in the space spanned by the columns of A can be expressed in a succinct manner by making use of the *augmented* matrix.

Definition 5.4
The matrix whose columns are $\mathbf{a}_1, \mathbf{a}_2, \ldots \mathbf{a}_n, \mathbf{b}$, denoted by $[A, \mathbf{b}]$, is called the **augmented matrix** of A.

We can now rephrase Theorem 5.2 into the following form:

Theorem 5.2A
The system of equations (5.6) has a solution if and only if $r(A) = r([A, \mathbf{b}])$.

We now come to the uniqueness problem. Assume we have a particular solution, say $\mathbf{y}$, to (5.6), that is,

$$A\mathbf{y} = \mathbf{b},$$

and that $\mathbf{z}$ is any solution to the corresponding system of homogeneous equations, to $A\mathbf{x} = \mathbf{O}$ so that

$$A\mathbf{z} = \mathbf{O}.$$

Next let $\mathbf{w} = \mathbf{y} + \mathbf{z}$, then

$$A\mathbf{w} = A(\mathbf{y} + \mathbf{z}) = A\mathbf{y} + A\mathbf{z} = \mathbf{b} + \mathbf{O} = \mathbf{b}$$

so that $\mathbf{w}$ is a solution to (5.6).

Conversely if $\mathbf{z}$ is a particular solution and $\mathbf{w}$ is another solution to (5.6), then

$$A\mathbf{z} = \mathbf{b} \text{ and } A\mathbf{w} = \mathbf{b}.$$

On subtracting, we find

$$A\mathbf{w} - A\mathbf{z} = A(\mathbf{w} - \mathbf{z}) = \mathbf{O},$$

that is,

$$A\mathbf{y} = \mathbf{O} \text{ where } \mathbf{y} = \mathbf{w} - \mathbf{z}, \text{ or } \mathbf{w} = \mathbf{z} + \mathbf{y}.$$

Theorem 5.3
If $\mathbf{w}$ is a solution to (5.6), then $\mathbf{w} = \mathbf{z} + \mathbf{y}$ where $\mathbf{z}$ is a particular solution to (5.6) and $\mathbf{y}$ is the solution to the corresponding homogeneous system.

Now to the number of solutions we can expect. If $r(A) \neq r([A,\mathbf{b}])$, the equations (5.6) are inconsistent, so that no solution exists. Assuming that $r(A) = r([A,\mathbf{b}])$ holds, we know (Theorem 5.2) that at least one solution to (5.6) exists.

The uniqueness of this solution is determined, as the consequence of Theorem 5.3, by the number of solutions to the corresponding homogeneous system $A\mathbf{x} = \mathbf{O}$.

We can summarise the results from the previous section, as follows:

Solutions to $A\mathbf{x} = \mathbf{b}$ exist provided that $r(A) = r([A,\mathbf{b}])$.

Case 1; $m = n$

(a) If $r(A) = n$, then A is non-singular and $A\mathbf{x} = \mathbf{O}$ has a unique solution ($\mathbf{x} = \mathbf{O}$). So (5.6) has the unique solution $\mathbf{x} = A^{-1}\mathbf{b}$.

(b) If $r(A) = r < n$,
$A\mathbf{x} = \mathbf{O}$ has $(n-r)$ linearly independent solutions.
So (5.6) has many solutions.

Case 2; $m < n$

If $r(A) = r$,
$A\mathbf{x} = \mathbf{O}$ has $(n-r)$ linearly independent solutions.
So (5.6) has many solutions.

In case 1(b) and case 2, we find the solutions as for homogeneous equations. We choose arbitrarily the values of the $(n-r)$ variables which then uniquely determine the remaining variables corresponding to the r linearly independent columns of A.

Example 5.2

Find the value of α for which the following system of equations is consistent and then find the solutions.

$$\begin{aligned} 2x + y - z - w &= 3 \\ -x + 2y + 3z + 3w &= 6 \\ x + 8y + 7z + 7w &= \alpha \,. \end{aligned}$$

Solution

We have seen (Ex. 5.1) that the rank of the matrix of the coefficients A is 2, and that the first two columns of A are linearly independent. For the equations to be consistent $r(A) = r([A,\mathbf{b}])$ so that $\mathbf{b}$ must be a linear combination of the first two columns, that is,

$$\begin{bmatrix} 3 \\ 6 \\ \alpha \end{bmatrix} = x_1 \begin{bmatrix} 2 \\ -1 \\ 1 \end{bmatrix} + x_2 \begin{bmatrix} 1 \\ 2 \\ 8 \end{bmatrix}.$$

On equating the first two components, we find

$$\begin{aligned} 3 &= 2x_1 + x_2 \\ 6 &= -x_1 + 2x_2, \end{aligned}$$

that is,

$$x_1 = 0,\ x_2 = 3.$$

Equating the third components

$$\alpha = x_1 + 8x_2 = 24.$$

We can now arbitrarily select values for z and w. Let $z = -2$ and $w = 3$, then

$$x\begin{bmatrix}2\\-1\\1\end{bmatrix} + y\begin{bmatrix}1\\2\\8\end{bmatrix} - 2\begin{bmatrix}-1\\3\\7\end{bmatrix} + 3\begin{bmatrix}-1\\3\\7\end{bmatrix} = \begin{bmatrix}3\\6\\24\end{bmatrix}$$

that is,

$$x\begin{bmatrix}2\\-1\\1\end{bmatrix} + y\begin{bmatrix}1\\2\\8\end{bmatrix} = \begin{bmatrix}4\\3\\17\end{bmatrix}, \text{ so that } x = 1 \text{ and } y = 2.$$

Hence a particular solution to the system of equations is $[1,2,-2,3]'$.

In Ex. 5.1 we found 2 linearly independent solutions to the homogeneous system, they are $[3,-3,1,2]'$ and $[1,-1,1,0]'$. It follows that every solution to the system of non-homogeneous equations has the form

$$[1,2,-2,3]' + \lambda[3,-3,1,2]' + \mu[1,-1,1,0]',$$

λ, μ being real numbers.

PROBLEMS FOR CHAPTER 5

1) Find the value of α for which the system

$$x - 2y + z - 3w = \alpha,$$

$$2x + y + z - 2w = 2,$$

$$7x - 4y + 5z - 13w = -8,$$

is consistent, and then find the solutions.

2) Determine the number of independent solutions for each of the systems of equations, and then find all the solutions.

(i)
$$\begin{aligned} x + y - z &= 0, \\ 2x - y - 5z &= 0, \\ 2x + 2y - 2z &= 0; \end{aligned}$$

(ii)
$$\begin{aligned} 2x + y - 4z &= 0, \\ x - y + z &= 0, \\ 4x - y - 2z &= 0; \end{aligned}$$

(iii)
$$\begin{aligned} x - y + z &= 0, \\ x - 3y + 5z &= 0, \\ 2x + y - 3z &= 0. \end{aligned}$$

3) Determine for each system of equations the rank of the matrix of coefficients A and the rank of the augmented matrix. Hence state whether the system has a solution.

(i)
$$\begin{aligned} x + 2y + 4z &= 3, \\ -x + y - z &= 0, \\ x - y + z &= 0; \end{aligned}$$

(ii)
$$\begin{aligned} x + 2y + 4z &= 5, \\ -x + y - z &= 1, \\ x - y + z &= 2. \end{aligned}$$

4) Three species of fish are introduced into a fish-pond. The fish are fed three types of food. Let $\mathbf{C}_i = (C_{i1}, C_{i2}, C_{i3})$ where C_{ij} is the average consumption per day of the jth type of food by a fish of the ith species ($i, j, = 1,2,3$). Suppose that

(i) $\mathbf{C}_1 = (2,1,2)$, $\mathbf{C}_2 = (3,2,1)$ and $\mathbf{C}_3 = (2,3,1)$.

(ii) $\mathbf{C}_1 = (2,1,2)$, $\mathbf{C}_2 = (3,2,1)$ and $\mathbf{C}_3 = (3,1,5)$

and that the supply is 4650, 4350, and 2500 units of food type 1, 2, and 3 respectively.

Assuming that all food is consumed, how many fish of each species can coexist in the pond?

(iii) The supply of food is altered to 5500, 2650, and 5900 units of type 1, 2, and 3 respectively. What is now the solution for (ii) above.

CHAPTER 6

Eigenvectors and Eigenvalues

In this chapter we shall be concerned with linear transformations of the type $T : U \to U$, of a vector space into itself. We shall define and use eigenvectors and eigenvalues to achieve a simplification of the matrix representations of the linear transformations.

We first consider vectors of type $\mathbf{x} \neq \mathbf{O}$, which under the transformation T maps to $\lambda\mathbf{x}$, λ being a scalar, that is, a vector whose direction is unchanged under T.

Definition 6.1

If a linear transformation T maps the vector $\mathbf{x}$ into $\lambda\mathbf{x}$, then $\mathbf{x}$ is called an **eigenvector** of T.

The scalar λ is called the corresponding **eigenvalue.**

$\mathbf{x}$ is also known as a **characteristic vector** or **latent vector** or **proper vector,** and λ as the corresponding **characteristic value, latent value** or **proper value.** We also refer to the eigenvalues and eigenvectors of a matrix A. The matrix A represents a linear transformation T relative to some fixed basis. We shall see (Theorem 6.1) that the choice of basis is immaterial, so far as the eigenvalues are concerned – they are invariant under similarity transformations.

Example 6.1

$\begin{bmatrix} 3 & -1 \\ 4 & -2 \end{bmatrix}$ represents a linear transformation, T, with respect to some basis. Verify that $[1,1]'$ is an eigenvector of T, and find the corresponding eigenvalue.

Solution

Since $$\begin{bmatrix} 3 & -1 \\ 4 & -2 \end{bmatrix}\begin{bmatrix} 1 \\ 1 \end{bmatrix} = \begin{bmatrix} 2 \\ 2 \end{bmatrix} = 2\begin{bmatrix} 1 \\ 1 \end{bmatrix}$$

hence $[1,1]'$ is an eigenvector and $\lambda = 2$ is the corresponding eigenvalue of T.

6.1 THE CHARACTERISTIC EQUATION

We shall consider that the linear transformation T is represented by the $n \times n$ matrix A with respect to some fixed basis, say the standard basis. To find the characteristic values and vectors of T, we need to find scalars λ, such that there exists $\mathbf{x} \neq \mathbf{O}$ satisfying

$$A\mathbf{x} = \lambda\mathbf{x} \quad \text{or} \quad [\lambda I - A]\,\mathbf{x} = \mathbf{O} \tag{6.1}$$

I being the unit matrix of order n. From Sec. 5.1 we know that (6.1) will have non-trivial solutions only if λ is such that $[\lambda I - A]$ is singular, that is, if $|\lambda I - A| = 0$, which, written out in full, is

$$\begin{vmatrix} \lambda - a_{11} & a_{12} & & a_{1n} \\ a_{21} & \lambda - a_{22} & \cdots & a_{2n} \\ \vdots & & & \\ a_{n1} & a_{n2} & & \lambda - a_{nn} \end{vmatrix} = 0. \tag{6.2}$$

On expansion we note that this is a polynomial, denoted by $c(\lambda)$, of degree n in λ. It has the form

$$c(\lambda) = \lambda^n + b_1\lambda^{n-1} + \ldots + b_{n-1}\lambda + b_n = 0. \tag{6.3}$$

(6.3) is called the **characteristic equation** of A, and its roots are the eigenvalues of A. To each eigenvalue of A there always exists at least one eigenvector $\mathbf{x}$.

Although for most purposes the above definition of an eigenvalue is acceptable, in a mathematically rigorous treatment we would need to specify that the roots of (6.3) are the eigenvalues of the matrix A only if they are 'scalars', that is, members of the field being considered. For example the complex (or purely imaginary) roots of (6.3) are not eigenvalues of A if the field of real numbers is being considered. In engineering applications, the field considered is nearly always complex, in which case all solutions of (6.3) are the eigenvalues of the matrix.

If $\lambda_1, \lambda_2, \ldots \lambda_n$ are the roots of this polynomial, we can write $c(\lambda)$ as

$$c(\lambda) = (\lambda - \lambda_1)(\lambda - \lambda_2) \ldots (\lambda - \lambda_n). \tag{6.4}$$

If we let $\lambda = 0$, we obtain

$$c(0) = (-1)^n \lambda_1 \lambda_2 \ldots \lambda_n = b_n \quad \text{(by 6.3)}.$$

Since $c(\lambda) = |\lambda I - A|$, $c(0) = |-A|$, then

$$|-A| = (-1)^n \lambda_1 \lambda_2 \ldots \lambda_n. \tag{6.5}$$

From this equation we note that if $|A| = 0$, some $\lambda_i = 0$, so a singular matrix has one or more zero eigenvalues, and conversely.
On expanding (6.4) we obtain

$$c(\lambda) = \lambda^n - (\lambda_1 + \lambda_2 + \ldots + \lambda_n)\, \lambda^{n-1} + \ldots,$$

Comparing the coefficient of λ^{n-1} in this equation and (6.3), we note that

$$b_1 = -(\lambda_1 + \lambda_2 + \ldots + \lambda_n).$$

By inspection of (6.2) we have

$$b_1 = -(a_{11} + a_{22} + \ldots + a_{nn}).$$

Definition 6.2
The sum of the diagonal elements $\sum_{i=1}^{n} a_{ii}$ of a square matrix A of order $n \times n$ is called the **trace** of the matrix, denoted by tr A. Hence

$$b_1 = -(\lambda_1 + \lambda_2 + \ldots + \lambda_n) = -\text{tr}\, A. \tag{6.6}$$

Example 6.2
(a) Find the characteristic equation and verify (6.6) for the matrix of Ex. 6.1.
(b) Find the eigenvectors.

Solution
(a)

$$|\lambda I - A| = \begin{vmatrix} \lambda-3 & 1 \\ -4 & \lambda+2 \end{vmatrix} = (\lambda-3)(\lambda+2) + 4$$

that is,

$$c(\lambda) = \lambda^2 - \lambda - 2 = (\lambda-2)(\lambda+1) = 0,$$

hence

$$\lambda_1 = 2, \lambda_2 = -1.$$

$$b_1 = -1, (\lambda_1 + \lambda_2) = 1, \text{tr}\, A = 3-2 = 1.$$

So (6.6) is verified.

(b) for $\lambda = 2$

$$[\lambda I - A]\mathbf{x}_1 = \mathbf{O} \text{ becomes } \begin{bmatrix} -1 & 1 \\ -4 & 4 \end{bmatrix}\begin{bmatrix} x_1 \\ x_2 \end{bmatrix} = \begin{bmatrix} 0 \\ 0 \end{bmatrix}.$$

On solving we find $\mathbf{x} = [1,1]'$ – as mentioned in Ex. 6.1, for $\lambda = -1$

$$[\lambda I - A]\mathbf{x}_2 = \mathbf{O} \text{ becomes } \begin{bmatrix} -4 & 1 \\ -4 & 1 \end{bmatrix}\begin{bmatrix} x_1 \\ x_2 \end{bmatrix} = \begin{bmatrix} 0 \\ 0 \end{bmatrix}.$$

On solving we find $\mathbf{x}_2 = [1,4]'$.

$\mathbf{x}_1$ and $\mathbf{x}_2$ are the eigenvectors corresponding to the two eigenvalues, $\lambda_1 = 2$ and $\lambda_2 = -1$ respectively. We note that they are linearly independent.

It will be interesting to examine whether eigenvalues, eigenvectors, and the characteristic equation change when the basis with respect to which the matrix representing the linear transformation T changes. We shall therefore assume that, relative to a new basis, T is represented by a matrix B, which (Eq. 3.18) is related to A by

$$B = P^{-1}AP.$$

Theorem 6.1

Given two similar (see definitions 3.6) matrices A and B, so that $B = P^{-1}AP$, then

(a) A and B have the same characteristic equation.

(b) A and B have the same eigenvalues.

(c) Given an eigenvalue λ and the corresponding eigenvector $\mathbf{x}$ of A, the corresponding eigenvector of B is $P^{-1}\mathbf{x}$.

Proof

(a) The characteristic equation of B is

$$\begin{aligned} \mathbf{O} = |\lambda I - B| &= |\lambda I - P^{-1}AP| = |\lambda P^{-1}IP - P^{-1}AP| \\ &= |P^{-1}||\lambda I - A||P| = |P^{-1}||P||\lambda I - A| \\ &= |\lambda I - A| = \text{characteristic equation of } A \end{aligned}$$

(since $|P^{-1}||P| = 1$).

(b) Follows from (a).

(c) Since $A\mathbf{x} = \lambda\mathbf{x}$, then

$$P^{-1}APP^{-1}\mathbf{x} = P^{-1}\lambda\mathbf{x},$$

that is,

$$B(P^{-1}\mathbf{x}) = \lambda(P^{-1}\mathbf{x})$$

which proves that $P^{-1}\mathbf{x}$ is the eigenvector corresponding to the eigenvalue λ of B.

Corollaries
If A has eigenvalues $\lambda_1,\lambda_2, \ldots \lambda_n$ then

(i) A^r has eigenvalues $\lambda_1^r,\lambda_2^r, \ldots \lambda_n^r$.
Indeed, since there exists a matrix P such that $P^{-1}AP = \Lambda = \text{diag}\{\lambda_1,\lambda_2, \ldots \lambda_n\}$, the eigenvalues of A^r are the eigenvalues of

$$\underbrace{(P^{-1}AP)(P^{-1}AP) \ldots (P^{-1}AP)}_{r \text{ factors}} = \Lambda\Lambda \ldots \Lambda = \Lambda^r$$

$$= \text{diag}\{\lambda_1^r,\lambda_2^r, \ldots \lambda_n^r\}.$$

(ii) provided that A is non-singular, the eigenvalues of A^{-1} are $\lambda_1^{-1},\lambda_2^{-1}, \ldots \lambda_n^{-1}$.
Indeed the eigenvalues of A^{-1} are the roots of the equation

$$|\mu I - A^{-1}| = 0.$$

Since $[\mu I - A^{-1}] = [\mu A - I]A^{-1} = \mu I[A - \mu^{-1}I]A^{-1}$, it follows that

$$|\mu I - A^{-1}| = (-1)^n \mu^n |A - \mu^{-1}I||A^{-1}|,$$

hence

$$|\mu I - A^{-1}| = 0$$

when

$$|\mu^{-1}I - A| = 0.$$

The result follows when this last equation is compared with Eqs (6.2) and (6.4).

Example 6.3

Given $A = \begin{bmatrix} 3 & -1 \\ 4 & -2 \end{bmatrix}$ and the matrix of transition $P = \begin{bmatrix} 2 & 3 \\ 1 & 2 \end{bmatrix}$, find

(a) $B = P^{-1}AP$.
(b) The characteristic equations of A and B.
(c) The eigenvalues and eigenvectors of A and B.

Solution

(a) $B = \begin{bmatrix} 2 & -3 \\ -1 & 2 \end{bmatrix}\begin{bmatrix} 3 & -1 \\ 4 & -2 \end{bmatrix}\begin{bmatrix} 2 & 3 \\ 1 & 2 \end{bmatrix} = \begin{bmatrix} -8 & -10 \\ 7 & 9 \end{bmatrix}$.

(b) The characteristic equation of A (see Ex. 6.2) is

$$c(\lambda) = \lambda^2 - \lambda - 2 = 0.$$

The characteristic equation of B is

$$\begin{vmatrix} \lambda+8 & 10 \\ -7 & \lambda-9 \end{vmatrix} = \lambda^2 - \lambda - 2 = 0.$$

(c) Eigenvalues of A and B are $\lambda_1 = 2$ and $\lambda_2 = -1$.
Eigenvectors of A (see Ex. 6.2) are $\mathbf{x}_1 = [1,1]'$ and $\mathbf{x}_2 = [1,4]'$.
For $\lambda = 2$

$$[\lambda_1 I - B]\mathbf{y}_1 = \mathbf{O} \text{ has solution } \mathbf{y}_1 = [-1,1]'.$$

For $\lambda = -1$

$$[\lambda_2 I - B]\mathbf{y}_2 = \mathbf{O} \text{ has solution } \mathbf{y}_2 = [-10,7]'.$$

Notice that

$$P^{-1}\mathbf{x}_1 = \begin{bmatrix} 2 & -3 \\ -1 & 2 \end{bmatrix}\begin{bmatrix} 1 \\ 1 \end{bmatrix} = \begin{bmatrix} -1 \\ 1 \end{bmatrix} = \mathbf{y}_1$$

and

$$P^{-1}\mathbf{x}_2 = \begin{bmatrix} 2 & -3 \\ -1 & 2 \end{bmatrix}\begin{bmatrix} 1 \\ 4 \end{bmatrix} = \begin{bmatrix} -10 \\ 7 \end{bmatrix} = \mathbf{y}_2 .$$

Theorem 6.1 proves that similar matrices have the same characteristic equations. The converse is false in general, as can be shown by inspecting the following two matrices:

$$A = \begin{bmatrix} 1 & 0 \\ 0 & 1 \end{bmatrix} \text{ and } B = \begin{bmatrix} 1 & 1 \\ 0 & 1 \end{bmatrix}.$$

Since for any non-singular matrix P of order 2×2, $P^{-1}AP = P^{-1}IP = P^{-1}P = A$, A and B are *not* similar, yet each has the characteristic equation $(\lambda-1)^2 = 0$.

6.2 THE EIGENVALUES OF THE TRANSPOSED MATRIX

Corresponding to the definition of the eigenvalues of A, that is, values of λ such that

$$A\mathbf{x} = \lambda\mathbf{x}$$

has a non-trivial solution, the eigenvalues of the transposed matrix A are defined by

$$A'\mathbf{y} = \mu\mathbf{y}.$$

Corresponding to (6.1) we have

$$[\mu I - A']\mathbf{y} = \mathbf{0} \tag{6.7}$$

which has non-trivial solutions when

$$|\mu I - A'| = 0.$$

Since $|A| = |A'|$, it follows that the eigenvalues for A are the same as those for A', that is, $\mu = \lambda$. On the other hand the eigenvectors are in general different; they are sometimes called **eigenrows**. If the eigenvectors corresponding to λ_i are $\mathbf{y}_i$ we have

$$A'\mathbf{y}_i = \lambda_i\mathbf{y}_i.$$

On taking the transpose, we find

$$\mathbf{y}_i'A = \lambda_i\mathbf{y}_i'.$$

This last equation defines a 'new' type of eigenvector of A. We sometimes distinguish it from the 'old' type by the following definition.

Definition 6.3
For an eigenvalue λ_i of A, the vectors $\mathbf{x}_i$ satisfying

$$A\mathbf{x}_i = \lambda_i \mathbf{x}_i$$

are called **right-hand** eigenvectors of A, whereas the vectors $\mathbf{y}_i$ such that

$$\mathbf{y}_i' A = \lambda_i \mathbf{y}_i'$$

are called **left-hand** eigenvectors or *eigenrows* of A corresponding to λ_i.

Theorem 6.2
If $\mathbf{x}_i$ is a right-hand eigenvector of A corresponding to an eigenvalue λ_i, and $\mathbf{y}_j$ is a left-hand eigenvector of A corresponding to an eigenvalue $\lambda_j (\lambda_i \neq \lambda_j)$, then

$$\mathbf{x}_i'\mathbf{y}_j = 0 \quad (i \neq j). \tag{6.8}$$

Proof
Since $A\mathbf{x}_i = \lambda_i \mathbf{x}_i$, on transposing we obtain

$$\mathbf{x}_i' A' = \lambda_i \mathbf{x}_i'.$$

Post-multiplying by $\mathbf{y}_j$, we find

$$\mathbf{x}_i' A' \mathbf{y}_j = \lambda_i \mathbf{x}_i' \mathbf{y}_j. \tag{6.9}$$

By definition,

$$A' \mathbf{y}_j = \lambda_j \mathbf{y}_j.$$

Pre-multiplying by $\mathbf{x}_i$, we find

$$\mathbf{x}_i' A' \mathbf{y}_j = \lambda_j \mathbf{x}_i' \mathbf{y}_j. \tag{6.10}$$

On subtracting (6.9) from (6.10), we have

$$(\lambda_j - \lambda_i)\mathbf{x}_i'\mathbf{y}_j = 0. \text{ The result follows.}$$

Example 6.4

Given $A = \begin{bmatrix} 2 & 1 \\ -5 & -2 \end{bmatrix}$

(a) Find the eigenvalues and the corresponding right-hand and left-hand eigenvectors.
(b) Verify Eq. (6.8) for the eigenvectors in (a).

Solution

(a) $|\lambda I - A| = \lambda^2 + 1 = 0$, hence $\lambda_1 = i$ and $\lambda_2 = -i$. First the right-hand eigenvectors.
For $\lambda_1 = i$

$$\begin{bmatrix} i-2 & -1 \\ 5 & i+2 \end{bmatrix} \begin{bmatrix} x_1 \\ x_2 \end{bmatrix} = \begin{bmatrix} 0 \\ 0 \end{bmatrix}, \text{ hence } \mathbf{x}_1 = \begin{bmatrix} x_1 \\ x_2 \end{bmatrix} = \begin{bmatrix} 1 \\ i-2 \end{bmatrix}.$$

For $\lambda_2 = -i$, we find $\mathbf{x}_2 = [1, -(i+2)]'$.
Next the left-hand eigenvectors.
For $\lambda_1 = i$, $[\lambda I - A']\mathbf{y} = \mathbf{0}$ becomes

$$\begin{bmatrix} i-2 & 5 \\ -1 & i+2 \end{bmatrix} \begin{bmatrix} y_1 \\ y_2 \end{bmatrix} = \begin{bmatrix} 0 \\ 0 \end{bmatrix} \text{ hence } \mathbf{y}_1 = \begin{bmatrix} 1 \\ \dfrac{-i+2}{5} \end{bmatrix},$$

For $\lambda_2 = -i$,

$$\begin{bmatrix} -i-2 & 5 \\ -1 & -i+2 \end{bmatrix} \begin{bmatrix} y_1 \\ y_2 \end{bmatrix} = \begin{bmatrix} 0 \\ 0 \end{bmatrix} \text{ hence } \mathbf{y}_2 = \begin{bmatrix} 1 \\ \dfrac{i+2}{5} \end{bmatrix}.$$

(b)

$$\mathbf{x}_1'\mathbf{y}_2 = [1, i-2] \begin{bmatrix} 1 \\ \dfrac{i+2}{5} \end{bmatrix} = 0, \quad \mathbf{x}_2'\mathbf{y}_1 = [1, -(i+2)] \begin{bmatrix} 1 \\ \dfrac{-i+2}{5} \end{bmatrix} = 0.$$

Notice that in the above example $\mathbf{x}_i'\mathbf{y}_i \neq 0$, $(i = 1,2)$, indeed

$$\mathbf{x}_1'\mathbf{y}_1 = [1, i-2] \begin{bmatrix} 1 \\ \frac{-i+2}{5} \end{bmatrix} = \frac{2+4i}{5} \text{ and } \mathbf{x}_2'\mathbf{y}_2 = [1, -(i+2)] \begin{bmatrix} 1 \\ \frac{i+2}{5} \end{bmatrix} = \frac{2-4i}{5}.$$

On the other hand, we know that if $\mathbf{x}$ is an eigenvector of a matrix, then any multiple, say $k\mathbf{x}$, is also an eigenvector of that matrix. Thus, for example,

$$\mathbf{z}_1 = \frac{5}{2+4i}\mathbf{y}_1 = \frac{1}{2}\begin{bmatrix} 1-2i \\ -i \end{bmatrix} \text{ and } \mathbf{z}_2 = \frac{5}{2-4i}\mathbf{y}_2 = \frac{1}{2}\begin{bmatrix} i+2i \\ i \end{bmatrix}$$

are right-hand eigenvectors of A corresponding to the eigenvalues $\lambda = i$ and $\lambda = -i$ respectively (as can be checked quite easily).

The relations (6.8) are satisfied (replacing the $\mathbf{y}_i$ by the $\mathbf{z}_i$), but

$$\mathbf{x}_i'\mathbf{z}_i = 1 \quad (i=1,2). \tag{6.11}$$

In general we can choose the right-hand and left-hand eigenvectors of a matrix so that conditions (6.11) hold; the relations (6.8) are always satisfied. Vectors satisfying (6.11) are said to be **normalized.**

6.3 WHEN ALL THE EIGENVALUES OF A ARE DISTINCT

The eigenvalues of an $n \times n$ matrix A are the roots of the polynomial equation of degree n

$$c(\lambda) = 0.$$

The roots can be real or complex, and distinct or multiple. Also, corresponding to an eigenvalue λ, there may be a single eigenvector or a set of linearly independent eigenvectors depending on the rank of the matrix

$$[\lambda I - A]$$

as discussed in Sec. 5.1.

Theorem 6.3
The eigenvectors of the matrix A of order $n \times n$ corresponding to distinct eigenvalues are

(a) linearly independent, and

(b) unique (save for a scalar multiple).

Proof
Let the distinct eigenvalues of A be $\lambda_1, \lambda_2, \ldots \lambda_n$, then $[\lambda_i I - A]$ $(i=1,2,\ldots n)$ is singular, and Eq. (6.1) has at least one solution $\mathbf{x}_i$ for each λ_i $(i=1,2,\ldots n)$.

Let $S = \{\mathbf{x}_1, \mathbf{x}_2, \ldots \mathbf{x}_n\}$ be the set of eigenvectors corresponding to the distinct eigenvalues. We prove by contradiction that the vectors in S are linearly independent.

Assume that there exists a *smallest* number $r \leq n$ of linearly dependent vectors in S, say $\mathbf{x}_1, \mathbf{x}_2, \ldots \mathbf{x}_r$. Then there are non-zero scalars $a_1, a_2, \ldots a_r$ such that

$$\sum_{i=1}^{r} a_i \mathbf{x}_i = \mathbf{0}. \tag{6.12}$$

Pre-multiplying (6.12) by A, we obtain

$$\sum_{i=1}^{r} a_i A \mathbf{x}_i = \sum_{i=1}^{r} a_i \lambda_i \mathbf{x}_i = \mathbf{0}. \tag{6.13}$$

Multiplying (6.12) by λ_r, we have

$$\sum_{i=1}^{r} a_i \lambda_r \mathbf{x}_i = 0. \tag{6.14}$$

On subtracting (6.13) from (6.14) we obtain

$$\sum_{i=1}^{r-1} a_i (\lambda_r - \lambda_i) \mathbf{x}_i = \mathbf{0}.$$

Since $\lambda_r \neq \lambda_i$ $(i \neq r)$, we must have

$$a_1 = a_2 = \ldots = a_{r-1} = 0 \quad \text{(hence } a_r = 0 \text{ also)},$$

contrary to our assumption. This proves part (a) of the theorem; the vectors in S are linearly independent and so constitute a basis for an n-dimensional vector-space V. Assume that there are two eigenvectors $\mathbf{x}_1$ and $\mathbf{y}_1$ corresponding to λ_1.

Since $\mathbf{x}_1, \mathbf{x}_2, \ldots \mathbf{x}_n$ span the space V, we can write

$$\mathbf{y}_1 = \sum_{i=1}^{n} a_i \mathbf{x}_i \tag{6.15}$$

where at least one of the scalars a_i is non-zero. Pre-multiplying (6.15) by A, we obtain

$$A\mathbf{y}_1 = \sum_{i=1}^{n} a_i A \mathbf{x}_i$$

that is,

$$\lambda_1 \mathbf{y}_1 = \sum_{i=1}^{n} a_i \lambda_i \mathbf{x}_i. \tag{6.16}$$

Multiplying (6.15) by λ_1 we obtain

$$\lambda_1 \mathbf{y}_1 = \sum_{i=1}^{n} a_i \lambda_1 \mathbf{x}_i. \tag{6.17}$$

Subtracting (6.17) from (6.16), we find

$$\mathbf{0} = \sum_{i=2}^{n} a_i (\lambda_i - \lambda_1) \mathbf{x}_i.$$

Since the $\mathbf{x}_i$ are independent and $\lambda_i - \lambda_1 \neq 0$ $(i \neq 1)$ we have

$$a_2 = a_3 = \ldots = a_n = 0.$$

But at least one of the a_i $(i=1,2,\ldots n)$ is non-zero, hence $a_1 \neq 0$. It follows that

$$\mathbf{y}_1 = a_1 \mathbf{x}_1$$

which proves part (b) of the theorem.

Note

By an analogous method we can prove that the left-hand eigenvectors of A; $\mathbf{y}_1, \mathbf{y}_2, \ldots \mathbf{y}_n$, that is the eigenvectors of A', are linearly independent (and unique). It follows that they also constitute a basis Q (say) for the n-dimensional vector space V.

We note that both the set S of right-hand eigenvectors and the set Q of left-hand eigenvectors of a matrix A having distinct eigenvalues, constitute bases for the vector space V. These vectors are related by (6.8) and can be normalised to satisfy (6.11). Such bases have a special name.

Definition 6.4
Two sets of bases $S = \{\mathbf{x}_1,\mathbf{x}_2, \ldots \mathbf{x}_n\}$ and $Q = \{\mathbf{y}_1,\mathbf{y}_2, \ldots \mathbf{y}_n\}$ for a vector space V are said to be **reciprocal** if

$$\mathbf{x}_i'\mathbf{y}_j = \mathbf{y}_j'\mathbf{x}_i = \delta_{ij} \; (i,j = 1,2, \ldots n) \tag{6.18}$$

where δ_{ij} is the Kronecker delta, that is,

$$\delta_{ij} = \begin{matrix} 1 \text{ when } i=j \\ 0 \text{ when } i \neq j. \end{matrix}$$

6.4 A REDUCTION TO A DIAGONAL FORM

The matrices X and Y whose columns are respectively the linearly independent eigenvectors of a matrix A, and A', play a very important role in the development of matrix theory.

Definition 6.5
Given the linearly independent eigenvectors $\mathbf{x}_1,\mathbf{x}_2, \ldots \mathbf{x}_n$ of A and $\mathbf{y}_1,\mathbf{y}_2, \ldots \mathbf{y}_n$ of A', the matrices

$$X = [\mathbf{x}_1,\mathbf{x}_2, \ldots \mathbf{x}_n] \text{ and } Y = [\mathbf{y}_1,\mathbf{y}_2, \ldots \mathbf{y}_n]$$

are respectively called the **modal matrix** of A and the **modal matrix** of A'. We can write the equations

$$A\mathbf{x}_i = \lambda_i\mathbf{x}_i \quad (i=1,2, \ldots n)$$

in matrix form as

$$AX = \Lambda X \tag{6.19}$$

where

$$\Lambda = \begin{bmatrix} \lambda_1 & 0 & 0 & \dots & 0 \\ 0 & \lambda_2 & 0 & \dots & 0 \\ \vdots & & & & \\ 0 & 0 & 0 & \dots & \lambda_n \end{bmatrix} = \text{diag}\{\lambda_1, \lambda_2, \dots \lambda_n\}$$ is called the **eigenvalue matrix** of A. Similarly

$$A'\mathbf{y}_i = \lambda_i \mathbf{y}_i \quad (i=1,2,\dots n)$$

is written in matrix form as

$$A'Y = \Lambda Y. \tag{6.20}$$

Since X is non-singular, (6.19) can be written as

$$X^{-1}AX = \Lambda. \tag{6.21}$$

If we use normalised eigenvectors, we make use of the relations (6.18) and obtain

$$Y'X = I \tag{6.22}$$

so that $Y' = X^{-1}$, that is, $Y = (X^{-1})'$. Notice from (6.21) that A and Λ are similar, hence

Theorem 6.4
The $n \times n$ matrix A is similar to its eigenvalue matrix Λ if the eigenvalues of A are distinct.

Notice that the condition that the eigenvalues of A are distinct, is a sufficient condition for A and Λ to be similar; on the other hand it is not a necessary condition. In Theorem 7.14 we discuss a necessary and sufficient condition.

Example 6.5

The matrix $A = \begin{bmatrix} 2 & 1 \\ -5 & -2 \end{bmatrix}$ was found to have (see Ex. 6.4) eigenvalues $\lambda_1 = i$ and $\lambda_2 = -i$. The corresponding normalised eigenvectors are $\mathbf{x}_1 = [1, i-2]'$, $\mathbf{x}_2 = [1, -(i+2)]'$ and $\mathbf{y}_1 = \frac{1}{2}[1-2i, -i]'$, $\mathbf{y}_2 = \frac{1}{2}[1+2i, i]'$. Verify (6.21).

Solution

$$X = \begin{bmatrix} 1 & 1 \\ i-2 & -(i+2) \end{bmatrix}.$$

We can invert X to find X^{-1}, or we can use the relation $X^{-1} = Y'$. Since

$$Y = \frac{1}{2}\begin{bmatrix} 1-2i & 1+2i \\ -i & i \end{bmatrix}, \qquad Y' = \frac{1}{2}\begin{bmatrix} 1-2i & -i \\ 1+2i & i \end{bmatrix},$$

hence

$$X^{-1}AX = \frac{1}{2}\begin{bmatrix} 1-2i & -i \\ 1+2i & i \end{bmatrix}\begin{bmatrix} 2 & 1 \\ -5 & -2 \end{bmatrix}\begin{bmatrix} 1 & 1 \\ i-2 & -(i+2) \end{bmatrix} = \begin{bmatrix} i & 0 \\ 0 & -i \end{bmatrix}.$$

Theorem 6.5
The eigenvalues of a Hermitian matrix are real, and the eigenvectors corresponding to distinct eigenvalues are orthoganal.

Proof
Consider an eigenvalue λ and the corresponding eigenvector $\mathbf{x}$ of a Hermitian matrix A,

$$A\mathbf{x} = \lambda\mathbf{x}. \tag{1}$$

Taking the conjugate transpose of (1) we obtain

$$\bar{\mathbf{x}}'A^* = \bar{\lambda}\bar{\mathbf{x}}'$$

that is,

$$\bar{\mathbf{x}}'A = \bar{\lambda}\bar{\mathbf{x}}' \quad (\text{since } A^* = A). \tag{2}$$

Post-multiplying be $\mathbf{x}$, (2) becomes

$$\bar{\mathbf{x}}'A\mathbf{x} = \bar{\lambda}\bar{\mathbf{x}}'\mathbf{x}. \tag{3}$$

Pre-multiplying by $\bar{\mathbf{x}}'$, (1) becomes

$$\bar{\mathbf{x}}'A\mathbf{x} = \lambda\bar{\mathbf{x}}'\mathbf{x}. \tag{4}$$

Comparing (3) and (4) we conclude that

$$\bar{\lambda}=\lambda, \text{ that is, } \lambda \text{ is real.}$$

Next consider the eigenvectors $\mathbf{x}_1$ and $\mathbf{x}_2$ corresponding to two distinct eigenvalues λ_1 and λ_2 of A, then

$$A\mathbf{x}_1 = \lambda_1\mathbf{x}_1, \quad A\mathbf{x}_2 = \lambda_2\mathbf{x}_2 \tag{5}$$

and

$$\mathbf{x}_2'A\mathbf{x}_1 = \lambda_1\bar{\mathbf{x}}_2'\mathbf{x}_1.$$

Taking the conjugate transpose of both sides

$$\bar{\mathbf{x}}_1'A\mathbf{x}_2 = \lambda_1\bar{\mathbf{x}}_1'\mathbf{x}_2. \tag{6}$$

Pre-multiplying the second equation of (5) by $\bar{\mathbf{x}}_1'$, we obtain

$$\bar{\mathbf{x}}_1'A\mathbf{x}_2 = \lambda_2\bar{\mathbf{x}}_1'\mathbf{x}_2. \tag{7}$$

Comparing (6) and (7), noting that $\lambda_1 \neq \lambda_2$, we conclude that

$$\bar{\mathbf{x}}_1'\mathbf{x}_2 = 0, \tag{8}$$

which shows that the two vectors are orthogonal.

Corollary

The above theorem applies as a special case when A is a real symmetric matrix. The eigenvalues λ_i are real, and so are (in this case) the corresponding eigenvectors $\mathbf{x}_i$. Equation (8) becomes

$$\mathbf{x}_1'\mathbf{x}_2 = 0.$$

We know (Theorem 6.3) that the eigenvectors are unique save for a scalar multiple, which we can choose to normalise the eigenvector.

The modal matrix X constructed from the normalised eigenvectors will therefore be orthogonal (or unitary in the case of a Hermitian matrix A).

Example 6.6
Find the orthogonal matrix X such that

$$X^{-1}AX \text{ is diagonal,}$$

$$\text{where } A = \begin{bmatrix} 1 & -2 & 7 \\ -2 & 10 & -2 \\ 7 & -2 & 1 \end{bmatrix}.$$

Solution
We find the eigenvalues of A by solving the characteristic equation

$$|\lambda I - A| = \lambda^3 - 12\lambda^2 - 36\lambda + 432 = 0$$

which has solutions $\lambda = -6, 6, 12$.

The eigenvector corresponding to $\lambda_1 = -6$ is $\mathbf{x} = [1, 0, -1]'$, hence the normalised eigenvector is

$$\mathbf{x}_1 = [\frac{1}{\sqrt{2}}, 0, -\frac{1}{\sqrt{2}}]'.$$

Similarly, the normalised eigenvectors corresponding to $\lambda_2 = 6$ and $\lambda_3 = 12$ are respectively

$$\mathbf{x}_2 = [\frac{1}{\sqrt{3}}, \frac{1}{\sqrt{3}}, \frac{1}{\sqrt{3}}]' \text{ and } \mathbf{x}_3 = [\frac{1}{\sqrt{6}}, -\frac{2}{\sqrt{6}}, \frac{1}{\sqrt{6}}]'$$

hence the modal matrix is

$$X = \begin{bmatrix} \frac{1}{\sqrt{2}} & \frac{1}{\sqrt{3}} & \frac{1}{\sqrt{6}} \\ 0 & \frac{1}{\sqrt{3}} & -\frac{2}{\sqrt{6}} \\ -\frac{1}{\sqrt{2}} & \frac{1}{\sqrt{3}} & \frac{1}{\sqrt{6}} \end{bmatrix}.$$

This is an orthoganal matrix and

$$X^{-1} = X' = \begin{bmatrix} \frac{1}{\sqrt{2}} & 0 & -\frac{1}{\sqrt{2}} \\ \frac{1}{\sqrt{3}} & \frac{1}{\sqrt{3}} & \frac{1}{\sqrt{3}} \\ -\frac{1}{\sqrt{6}} & -\frac{2}{\sqrt{6}} & \frac{1}{\sqrt{6}} \end{bmatrix}.$$

It is tedious, but not difficult to verify that

$$X'AX = \Lambda = \text{diag}\{-6, 6, 12\}.$$

6.5 MULTIPLE EIGENVALUES

We have up till now assumed that the characteristic equation $c(\lambda) = 0$ of a matrix A has distinct roots. We now consider the case when c may have multiple roots.

Definition 6.6
When $(\lambda-\lambda_s)^k$ is a factor of $c(\lambda) = 0$ but $(\lambda-\lambda_s)^{k+1}$ is not, then $\lambda = \lambda_s$ is said to be a root of $c(\lambda) = 0$ of **algebraic multiplicity** k.

Corresponding to an eigenvalue λ of A there will correspond either one or more linearly independent eigenvectors.

Theorem 6.6
Consider the matrix A representing the linear transformation T: $V \to V$. If W is the set of all eigenvectors of A corresponding to an eigenvalue λ, together with the zero vector, then W is a subspace of V.

Proof
Let $\mathbf{x}_1$ and $\mathbf{x}_2 \in W$, and α and β be scalars, then $A\mathbf{x}_1 = \lambda\mathbf{x}_1$ and $A\mathbf{x}_2 = \lambda\mathbf{x}_2$ implies that

$$\begin{aligned} A(\alpha\mathbf{x}_1 + \beta\mathbf{x}_2) &= A(\alpha\mathbf{x}_1) + A(\beta\mathbf{x}_2) \\ &= \alpha A\mathbf{x}_1 + \beta A\mathbf{x}_2 = \alpha\lambda\mathbf{x}_1 + \beta\lambda\mathbf{x}_2 \\ &= \lambda(\alpha\mathbf{x}_1 + \beta\mathbf{x}_2). \end{aligned}$$

Hence $\alpha\mathbf{x}_1 + \beta\mathbf{x}_2 \in W$ and W is a subspace.
Because of its importance the space W has a special name.

Definition 6.7
The vector space W is called an **eigenspace** of A corresponding to λ.

Our aim is to find, if it exists, a basis in V with respect to which the linear transformation T has a diagonal matrix representation. We have seen in the last section that we can achieve this if we can construct a modal matrix from the linearly independent eigenvectors of A, the matrix representing T relative to an arbitrary basis for V. But in the case of an eigenvalue λ of multiplicity $k > 1$, there is no guarantee that the number of linearly independent eigenvectors corresponding to λ_s (equal to the dimension of W) will be sufficient to form a modal matrix.

Theorem 6.7
The dimension of an eigenspace W corresponding to λ_s does not exceed the algebraic multiplicity of λ_s.

Proof
If we let

$$[\lambda_s I - A] = B,$$

then the eigenvectors of A corresponding to λ_s are seen to be the solutions to

$$B\mathbf{x} = \mathbf{0}, \tag{6.23}$$

so that the dimension of W is $(n - r)$ (see Theorem 5.1) where n is the dimension of the space V and r is the rank of B.

The problem is, of course, that in general we do not know the rank of B.

On the other hand, if we assume that there are m linearly independent solutions to (6.23) say $\{\mathbf{w}_1, \mathbf{w}_2, \ldots \mathbf{w}_m\}$, they form a basis for W, which can be extended (Theorem 2.7) to a basis $S = \{\mathbf{w}_1, \mathbf{w}_2, \ldots, \mathbf{w}_m, \ldots \mathbf{w}_n\}$ for V.

We now make use of the result proved in Theorem 6.1 that the characteristic equation for A and for $P^{-1}AP$ (where P is any non-singular matrix) is the same, that is, it is independent of the choice of a basis for V.

Choosing the particular basis S for V, we have by definition

$$T(\mathbf{w}_i) = \lambda_s \mathbf{w}_i \quad (i = 1, 2, \ldots m)$$

and for $i > m$, we can express the image vectors $T(\mathbf{w}_i)$ as linear combinations of the basis vectors, that is;

$$T(\mathbf{w}_i) = a_{1i}\mathbf{w}_1 + a_{2i}\mathbf{w}_2 + \ldots + a_{ni}\mathbf{w}_n \quad (i = m+1, \ldots n).$$

Hence the matrix representing T with respect to S is (see Sec. 3.3)

$$H = \begin{bmatrix} \lambda_s & 0 & \dots & 0 & a_{1,m+1} & \dots & a_{1,n} \\ 0 & \lambda_s & & 0 & a_{2,m+1} & \dots & a_{2,n} \\ \vdots & \vdots & & \vdots & \vdots & & \vdots \\ 0 & 0 & & \lambda_s & a_{m,m+1} & \dots & a_{m,n} \\ \vdots & \vdots & & \vdots & \vdots & & \vdots \\ 0 & 0 & & 0 & a_{n,m+1} & & a_{n,n} \end{bmatrix} \leftarrow m^{\text{th}} \text{ row}$$

$\uparrow$ m^{th} column $\quad$ $\uparrow$ $(m+1)^{\text{st}}$ column

We can write H as a partitioned matrix

$$H = \left[\begin{array}{c|c} \lambda_s I_m & G \\ \hline O_{n-m,m} & D \end{array}\right]$$

where I_m is the unit matrix of order $m \times m$.

$O_{n-m,m}$ is the zero matrix of order $(n-m) \times m$.

$$G = \begin{bmatrix} a_{1,m+1} & \dots & a_{1,n} \\ \vdots & & \\ a_{m,m+1} & \dots & a_{m,n} \end{bmatrix} \text{ and } D = \begin{bmatrix} a_{m+1,m+1} & \dots & a_{m+1,n} \\ \vdots & & \\ a_{n,m+1} & \dots & a_{n,n} \end{bmatrix}.$$

Hence

$$c(\lambda) = |\lambda I - H| = \left|\begin{array}{c|c} \lambda I_m - \lambda_s I_m & G \\ \hline 0 & \lambda I_{n-m} - D \end{array}\right| = (\lambda - \lambda_s)^m \, g(\lambda) \quad (6.24)$$

where $g(\lambda) = |\lambda I_m - D|$.

Having assumed that the dimension of W is m, and taking into account the fact that g may itself have a factor $(\lambda-\lambda_s)^q$ $(q = 1,2,\ldots n-m)$, (6.24) shows that the multiplicity of λ_s is at *least* m. This proves the theorem.

Example 6.7

Discuss the result of Theorem 6.7 in terms of the following matrices:

(a) and (b)

$$A = \begin{bmatrix} 3 & 1 \\ 0 & 3 \end{bmatrix} \qquad A = \begin{bmatrix} 1 & 0 & 0 \\ -1 & 0 & 0 \\ 1 & 1 & 1 \end{bmatrix}.$$

Solution

(a) $c(\lambda) = |\lambda I - A| = (\lambda-3)^2 = 0$ for $\lambda_1 = \lambda_2 = 3$. Hence $\lambda_1 = 3$ is an eigenvalue of multiplicity 2. Using the notation of the above section, we have

$$B = [\lambda_1 I - A] = \begin{bmatrix} 0 & -1 \\ 0 & 0 \end{bmatrix}.$$

Since $r(B) = 1$, dimension $W = n-r = 1$, hence we expect only one (linearly independent) eigenvector corresponding $\lambda_1 = 3$.

Indeed $B\mathbf{x} = \mathbf{O}$ becomes $$\begin{bmatrix} 0 & -1 \\ 0 & 0 \end{bmatrix}\begin{bmatrix} x_1 \\ x_2 \end{bmatrix} = \begin{bmatrix} 0 \\ 0 \end{bmatrix}$$

so that $\mathbf{x}_1 = [1, 0]'$.

In this case we note that

$$\text{dimension } W\,(=1) < \text{multiplicity of } \lambda_s\,(=2)$$

satisfying the result of Theorem 6.7.

(b) $c(\lambda) = \lambda(\lambda-1)^2 = 0$ for $\lambda_1 = 0$, and $\lambda_2 = \lambda_3 = 1$.

For $\lambda_1 = 0$

$$B\mathbf{x}_1 = [\lambda_1 I - A]\,\mathbf{x}_1 = \begin{bmatrix} -1 & 0 & 0 \\ 1 & 0 & 0 \\ -1 & -1 & -1 \end{bmatrix} \begin{bmatrix} x_1 \\ x_2 \\ x_3 \end{bmatrix} = \begin{bmatrix} 0 \\ 0 \\ 0 \end{bmatrix}.$$

On solving we find the eigenvector $\mathbf{x}_1 = [0, 1, -1]'$.

For $\lambda_2 = 1$

$$B\mathbf{x} = \begin{bmatrix} 0 & 0 & 0 \\ 1 & 1 & 0 \\ -1 & -1 & 0 \end{bmatrix} \begin{bmatrix} x_1 \\ x_2 \\ x_3 \end{bmatrix}.$$

We note that since $r(B) = 1$, dimension $W = 3 - 1 = 2$, hence we expect *two* linearly independent eigenvectors corresponding to $\lambda = 1$. Indeed $B\mathbf{x} = \mathbf{0}$ has solutions $\mathbf{x}_2 = [1, -1, 0]'$ and $\mathbf{x}_3 = [0, 0, 1]'$. In this case dimension $W(=2)$ = multiplicity of $\lambda_s(=2)$ satisfying the result of Theorem 6.7.

In this case we can form a modal matrix (since we have 3 linearly independent eigenvectors) which will reduce A to a diagonal form. The modal matrix is

$$X = \begin{bmatrix} 0 & 1 & 0 \\ 1 & -1 & 0 \\ -1 & 0 & 1 \end{bmatrix}, \text{ so that } X^{-1} = \begin{bmatrix} 1 & 1 & 0 \\ 1 & 0 & 0 \\ 1 & 1 & 1 \end{bmatrix}$$

and we verify that

$$X^{-1}AX = \begin{bmatrix} 1 & 1 & 0 \\ 1 & 0 & 0 \\ 1 & 1 & 1 \end{bmatrix} \begin{bmatrix} 1 & 0 & 0 \\ -1 & 0 & 0 \\ 1 & 1 & 1 \end{bmatrix} \begin{bmatrix} 0 & 1 & 0 \\ 1 & -1 & 0 \\ -1 & 0 & 1 \end{bmatrix} = \begin{bmatrix} 0 & 0 & 0 \\ 0 & 1 & 0 \\ 0 & 0 & 1 \end{bmatrix}$$

$$= \Lambda = \text{diag}\,\{0, 1, 1\}.$$

Notice that the order of the eigenvalues in Λ, that is, 0, 1, 1, corresponds to the order of the eigenvectors used to form the modal matrix X, that is, $\mathbf{x}_1, \mathbf{x}_2, \mathbf{x}_3$.

6.6 THE CAYLEY-HAMILTON THEOREM

As we have seen, operations with scalars are, in general, different from operations with matrices.

For example, if

$xy = o$, then $x = o$ or $y = o$,

but if

$AB = 0$, it does not follow that either $A = 0$ or $B = 0$ (see Ex. 1.10).

If $ax = ay$ $(a \neq o)$ then $x = y$, but, if

$AX = AY$ $(A \neq 0)$, it does not follow that $X = Y$. (See Ex. 1.10).

On the other hand we have defined various powers of matrices and have seen that for a square matrix A

$$A^r A^s = A^{r+s} = A^s A^r \quad \text{(see Def. 1.10).}$$

It would seem that we could define a polynomial in A in a manner similar to the definition of a polynomial in the scalar variable λ (see Ex. 1.11)

Definition 6.8

Corresponding to a polynomial in a scalar variable λ

$$f(\lambda) = \lambda^k + b_1\lambda^{k-1} + \ldots + b_{k-1}\lambda + b_k,$$

we define the (square) matrix $f(A)$, called a **matrix polynomial**, by

$$f(A) = A^k + b_1 A^{k-1} + \ldots + b_{k-1}A + b_k I \tag{6.25}$$

where A is a matrix of order $n \times n$.

A number of properties of such polynomials are easily proved. For example,

I. If $f(\lambda), g(\lambda)$ are two polynomials such that

$$f(\lambda) = g(\lambda) \quad (\text{all } \lambda)$$

then

$$f(A) = g(A),$$

and

II. since $f(\lambda)g(\lambda) = g(\lambda)f(\lambda)$, then

$$f(A)g(A) = g(A)f(A),$$

hence, if $g(A) \neq 0$, then

$$[g(A)]^{-1}f(A)g(A)[g(A)]^{-1} = [g(A)]^{-1}g(A)f(A)[g(A)]^{-1},$$

that is,

$$[g(A)]^{-1}f(A) = f(A)[g(A)]^{-1}.$$

Example 6.8
Evaluate the matrix polynomial $f(A)$ corresponding to

$$f(\lambda) = \lambda^2 - 2\lambda + 3$$

given $A = \begin{bmatrix} 1 & 1 \\ -1 & 3 \end{bmatrix}$.

Solution

$$f(A) = A^2 - 2A + 3I = \begin{bmatrix} 0 & 4 \\ -4 & 8 \end{bmatrix} - 2\begin{bmatrix} 1 & 1 \\ -1 & 3 \end{bmatrix} + 3\begin{bmatrix} 1 & 0 \\ 0 & 1 \end{bmatrix}$$

$$= \begin{bmatrix} 0-2+3 & 4-2+0 \\ -4+2+0 & 8-6+3 \end{bmatrix} = \begin{bmatrix} 1 & 2 \\ -2 & 5 \end{bmatrix}.$$

We are particularly interested in polynomials f of degree k (say) having the property that $f(A) = 0$, which will allow us to express A^k in terms of lower powers of A.

In Sec. 2.1 it was mentioned that the set of all matrices over a field F of order $n \times n$ are elements of a vector space V of dimension n^2.

It must follow that the $(n^2 + 1)$ elements in the set $S = \{I, A, A^2, \ldots A^{n^2}\}$ are linearly dependent, that is, there must be a polynomial

$$f(\lambda) = a_0 + a_1\lambda + \ldots + a_{n^2}\lambda^{n^2}$$

such that

$$f(A) = 0.$$

Example 6.9

Given $f(\lambda) = 4 + \lambda^2 - 3\lambda^3 + \lambda^4$, and $A = \begin{bmatrix} 1 & 1 \\ -1 & 3 \end{bmatrix}$, verify that $f(A) = 0$.

Solution

Note that A is of order 2×2, hence $n^2 = 4$.

$$A = \begin{bmatrix} 1 & 1 \\ -1 & 3 \end{bmatrix}, \; A^2 = \begin{bmatrix} 0 & 4 \\ -4 & 8 \end{bmatrix}, \; A^3 = \begin{bmatrix} -4 & 12 \\ -12 & 20 \end{bmatrix}, \; A^4 = \begin{bmatrix} -16 & 32 \\ -32 & 48 \end{bmatrix}.$$

Hence

$$f(A) = \begin{bmatrix} 4 & 0 \\ 0 & 4 \end{bmatrix} + \begin{bmatrix} 0 & 4 \\ -4 & 8 \end{bmatrix} - \begin{bmatrix} -12 & 36 \\ -36 & 60 \end{bmatrix} + \begin{bmatrix} -16 & 32 \\ -32 & 48 \end{bmatrix} = \begin{bmatrix} 0 & 0 \\ 0 & 0 \end{bmatrix}.$$

Having established that there exists a polynomial f of degree n^2 such that $f(A) = 0$ if A is of order $n\times n$, we next ask if there exist other polynomials of degree $\leq n^2$ which have this property.

We shall prove that every square matrix satisfies its own characteristic equations. First, an example.

Example 6.10

Find the characteristic equation of the matrix A of Ex. 6.9. Show that A satisfies this equation.

Solution

$$C(\lambda) = |\lambda I - A| = \begin{vmatrix} \lambda-1 & -1 \\ 1 & \lambda-3 \end{vmatrix} = \lambda^2 - 4\lambda + 4 = 0$$

$$C(A) = A^2 - 4A + 4I = \begin{bmatrix} 0 & 4 \\ -4 & 8 \end{bmatrix} - \begin{bmatrix} 4 & 4 \\ -4 & 12 \end{bmatrix} + \begin{bmatrix} 4 & 0 \\ 0 & 4 \end{bmatrix} = \begin{bmatrix} 0 & 0 \\ 0 & 0 \end{bmatrix}.$$

In this case we have found a polynomial C of degree 2 which has the desired property.

To prove this result in general we need to understand that if we have a matrix $B(\lambda)$ whose elements are polynomials of degree $\leq n$ in λ, then we can write it as a matrix polynomial:

$$B(\lambda) = B_0 + B_1\lambda + B_2\lambda^2 + \ldots + B_n\lambda^n$$

where the elements of B_i are independent of λ.

Example 6.11
Write the matrix

$$B(\lambda) = \begin{bmatrix} \lambda^3-2\lambda+3 & \lambda^4+3\lambda^3 \\ \lambda-5 & \lambda^2+2\lambda+3 \end{bmatrix}$$

as a matrix polynomial.

Solution

$$B(\lambda) = \begin{bmatrix} 0 & 1 \\ 0 & 0 \end{bmatrix}\lambda^4 + \begin{bmatrix} 1 & 3 \\ 0 & 0 \end{bmatrix}\lambda^3 + \begin{bmatrix} 0 & 0 \\ 0 & 1 \end{bmatrix}\lambda^2 + \begin{bmatrix} -2 & 0 \\ 1 & 2 \end{bmatrix}\lambda + \begin{bmatrix} 3 & 0 \\ -5 & 3 \end{bmatrix}.$$

Theorem 6.8
The Cayley-Hamilton Theorem

Every square matrix A satisfies its own characteristic equation.

Proof
Let A be a matrix of order $n \times n$. The characteristic equation of A is

$$C(\lambda) = |\lambda I - A| = \lambda^n + b_1\lambda^{n-1} + \ldots b_{n-1}\lambda + b_n = 0. \quad (6.26)$$

By Def. 4.7, adj $[\lambda I - A] = B(\lambda)$ (say) is a matrix of co-factors of $[\lambda I - A]$. By inspection of (6.2) we note that each co-factor is a polynomial of degree $\leq (n-1)$ in λ. Hence $B(\lambda)$ is a matrix whose elements are polynomials of degree $\leq (n-1)$ in λ and (as in Ex. 6.11) can be written in the form

$$B(\lambda) = B_0 + B_1\lambda + \ldots + B_{n-1}\lambda^{n-1}. \quad (6.27)$$

By Theorem 4.11 we have

$$[\lambda I - A] \text{ adj } [\lambda I - A] = |\lambda I - A| I$$

that is,

$$[\lambda I - A]\ [B_0 + B_1\lambda + \ldots B_{n-1}\lambda^{n-1}] = \lambda^n I + b_1\lambda^{n-1} I + \ldots + b_n I.$$

Equating the coefficients of the various powers of λ we obtain:

$$\begin{aligned}
\lambda^n &: B_{n-1} = I \\
\lambda^{n-1} &: B_{n-2} - AB_{n-1} = b_1 I \\
\lambda^{n-2} &: B_{n-3} - AB_{n-2} = b_2 I \\
&\vdots \\
\lambda^1 &: B_0 - AB_1 = b_{n-1} I \\
\lambda^0 &: -AB_0 = b_n I.
\end{aligned}$$

Pre-multiplying the above equations by $A^n, A^{n-1}, \ldots A, I$ respectively, we obtain

$$\begin{aligned}
A^n B_{n-1} &= A^n \\
A^{n-1}B_{n-2} - A^n B_{n-1} &= b_1 A^{n-1} \\
A^{n-2}B_{n-3} - A^{n-1}B_{n-2} &= b_2 A^{n-2} \\
&\vdots \\
AB_0 - A^2 B_1 &= b_{n-1} A \\
-AB_0 &= b_n I\ .
\end{aligned}$$

Adding these equations, we find

$$0 = A^n + b_1 A^{n-1} + b_2 A^{n-2} + \ldots + b_{n-1}A + b_n I$$

that is, $0 = C(A)$, which proves the theorem.

Example 6.12

Given $A = \begin{bmatrix} 1 & 1 \\ -1 & 3 \end{bmatrix}$, use the Cayley-Hamilton theorem to evaluate A^5.

Solution
We have found (Ex. 6.9) the characteristic equation

$$C(\lambda) = \lambda^2 - 4\lambda + 4 = 0$$

that is, $C(A) = A^2 - 4A + 4I = 0$, so that $A^2 = 4A - 4I$.

Now

$$\begin{aligned} A^5 = A^3A^2 = A^3(4A-4I) &= 4A^4 - 4A^3 \\ &= 4A^2(4A-4I) - 4A^3 \\ &= 12A^3 - 16A^2 \\ &= 12A(4A-4I) - 16A^2 \\ &= 32A^2 - 48A \\ &= 32(4A-4I) - 48A \\ &= 80A - 128I \end{aligned}$$

$$= \begin{bmatrix} 80 & 80 \\ -80 & 240 \end{bmatrix} - \begin{bmatrix} 128 & 0 \\ 0 & 128 \end{bmatrix} = \begin{bmatrix} -48 & 80 \\ -80 & 112 \end{bmatrix}.$$

The characteristic polynomial C is not necessarily the polynomial of lowest degree such that $C(A) = 0$.

Definition 6.9
The polynomial of lowest degree $m(\lambda)$ whose coefficient of the highest power of λ is unity (that is, a **monic** polynomial) and such that $m(A) = 0$, where A is a matrix of order $n \times n$, is called the **minimum** (or **minimal**) **polynomial** of A.

One procedure for evaluating the minimum polynomial of a matrix A is the following routine:

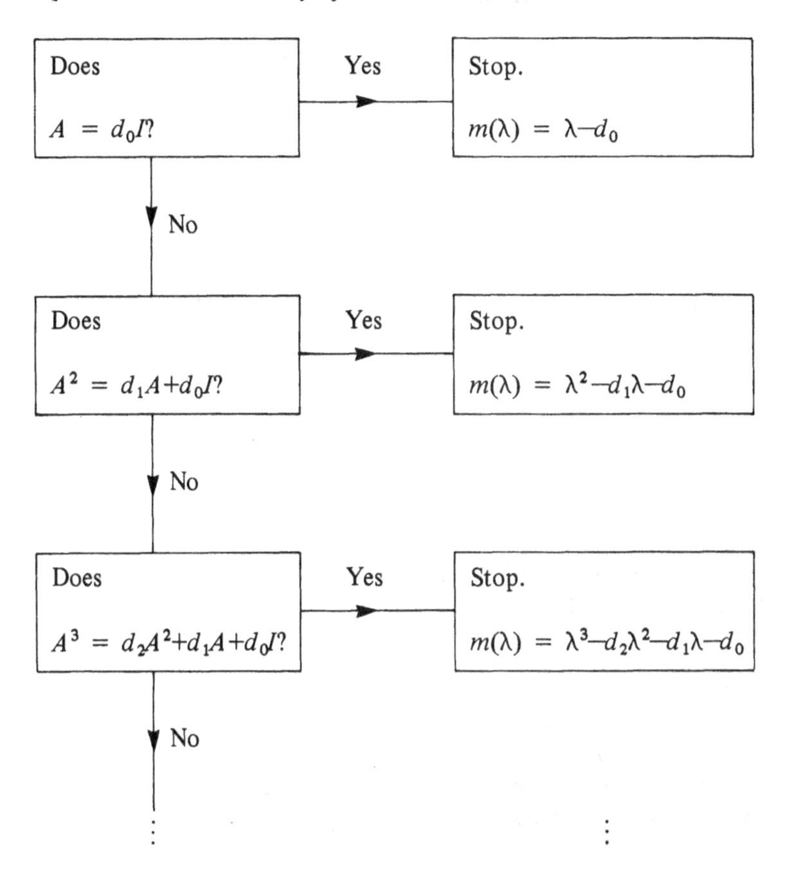

Example 6.13
Find the minimum polynomial of

$$A = \begin{bmatrix} 1 & 0 & 0 \\ -1 & 0 & 0 \\ 1 & 1 & 1 \end{bmatrix}.$$

Solution
In Ex. 6.7 part (b), we found that the characteristic equation of A is $\lambda^3 - 2\lambda^2 + \lambda = 0$. Therefore we know that $A^3 - 2A^2 + A = 0$. But is this the minimum polynomial? We use the above procedure.

Obviously $A \neq d_0 I$.

If $A^2 = d_1 A + d_0$
that is,

$$\begin{bmatrix} 1 & 0 & 0 \\ -1 & 0 & 0 \\ 1 & 1 & 1 \end{bmatrix} = d_1 \begin{bmatrix} 1 & 0 & 0 \\ -1 & 0 & 0 \\ 1 & 1 & 1 \end{bmatrix} + d_0 \begin{bmatrix} 1 & 0 & 0 \\ 0 & 1 & 0 \\ 0 & 0 & 1 \end{bmatrix}$$

then $1 = d_1 + d_0$

$-1 = -d_1$ hence $d_0 = 0$ and $d_1 = 1$.

It follows that $m(\lambda) = \lambda^2 - \lambda = \lambda(\lambda - 1)$.
In this case $A^2 - A = 0$.

Definition 6.10
Given a matrix A of order $n \times n$, any polynomial f such that $f(A) = 0$ is called an **annihilating polynomial** for A.

In particular, both the characteristic polynomial and the minimal polynomial are annihilating polynomials.

PROBLEMS FOR CHAPTER 6

1) Given the symmetric matrix

$$A = \begin{bmatrix} \frac{3}{2} & -\frac{1}{2} \\ -\frac{1}{2} & \frac{3}{2} \end{bmatrix}$$

find the modal matrix X such that $X^{-1}AX$ is diagonal. Show that X is orthogonal.

2) Verify that the eigenvectors of the matrix

$$A = \begin{bmatrix} 6 & 2 \\ 2 & 3 \end{bmatrix}$$

are orthogonal.
Find the matrix P such that $P^{-1}AP$ is diagonal.

3) Find an orthonormal set of eigenvectors for the matrix

$$A = \begin{bmatrix} 3 & 2 & 2 \\ 2 & 2 & 0 \\ 2 & 0 & 4 \end{bmatrix}.$$

4) Show that the eigenvalues of the Hermitian matrix

$$A = \begin{bmatrix} 1 & 1+2i \\ 1-2i & 5 \end{bmatrix}$$

are real and that the eigenvectors are orthogonal.

5) Find the eigenvalues of the Pauli matrices (see Problem 11, Chapter 1);

$$X_1 = \frac{1}{2}\begin{bmatrix} 0 & 1 \\ 1 & 0 \end{bmatrix}, \quad X_2 = \frac{1}{2}\begin{bmatrix} 0 & -i \\ i & 0 \end{bmatrix} \text{ and } X_3 = \frac{1}{2}\begin{bmatrix} 1 & 0 \\ 0 & -1 \end{bmatrix}.$$

6) Find the characteristic equation and the trace for each of the following matrices:

(i) $$A = \begin{bmatrix} 1 & 2 & 1 \\ 2 & 0 & -2 \\ -1 & 2 & 3 \end{bmatrix};$$

(ii)

$$A = \begin{bmatrix} 3 & 0 & 4 \\ 0 & 5 & 0 \\ 4 & 0 & 3 \end{bmatrix};$$

(iii)

$$A = \begin{bmatrix} 0 & 1 & 0 & 0 \\ 0 & 0 & 1 & 0 \\ 0 & 0 & 0 & 1 \\ -a_4 & -a_3 & -a_2 & -a_1 \end{bmatrix}.$$

7) Find the normalised column and row eigenvectors for the matrices

(i)

$$A = \begin{bmatrix} 2 & 1 \\ 2 & 3 \end{bmatrix};$$

(ii)

$$A = \begin{bmatrix} 3 & 4 \\ 4 & -3 \end{bmatrix}.$$

8) Find the modal matrices X and Y for the matrices A of Problem 7 above, using the normalised eigenvectors.

Check in the case of the symmetric matrix (ii) that these matrices are orthogonal, that is, that

$$X'X = I = Y'Y.$$

Verify that $X^{-1}AX$ and $Y^{-1}A'Y$ are diagonal matrices.

9) Determine a basis for, and the dimension of, the eigenspace associated with each eigenvalue of

(i)
$$A = \begin{bmatrix} 7 & 4 & -4 \\ 4 & 7 & -4 \\ -1 & -1 & 4 \end{bmatrix};$$

(ii)
$$A = \begin{bmatrix} 0 & 1 & 0 \\ 1 & 0 & 0 \\ 0 & 0 & 1 \end{bmatrix}.$$

10) Determine the minimal polynomial of A for each matrix in Problem 9 above. Hence evaluate A^4.

CHAPTER 7

Canonical Forms and Matrix Functions

In the last chapter we saw that under a similarity transformation we can reduce a matrix A of order $n \times n$ to a diagonal form if the dimension of the eigenspace W_i corresponding to each eigenvalue λ_i of A equals the algebraic multiplicity of λ_i.

The problem we shall investigate in this chapter is the reduction of A to a 'simple' form when the dimension of W_i is smaller than the algebraic multiplicity of λ_i. We shall discuss what is meant by 'simple' later on in this chapter.

In Sec. 7.1 we review very briefly certain aspects of polynomial functions theory which will be found useful in the development of this chapter. We do not include proofs of the various theorems; they can be found in various books on abstract algebra (for example see Refs. 1, 11, and 15).

7.1 POLYNOMIALS

Any expression of the form

$$f : x \to a_0 + a_1 x + \ldots + a_n x^n$$

with coefficients $a_i \epsilon F$ is called a **polynomial in** x **over** F. When the leading coefficient $a_n \neq 0$, we say that f is of **degree** n. If $a_n = 1$, f is said to be a *monic* polynomial (as already defined).

Theorem 7.1 (The Division Algorithm)
Let f and g be polynomials and $g(x) \neq 0$, then there exist polynomial q (the **quotient**) and r (the **remainder**) such that

$$f(x) = q(x)g(x) + r(x)$$

where the degree of $r <$ degree g.
We write the last statement as $\deg r < \deg g$. If $r(x) = 0$, then $f(x) = q(x)g(x)$, and g is said to be a **divisor** of f, and f is a **multiple** of g.

Definition 7.1
Any two polynomials f and g related by an equation

$$f(x) = ag(x), a \epsilon F \qquad \text{(that is, } a \text{ being a constant)}$$

are called **associates**.

Definition 7.2
A polynomial f is said to be **irreducible** or **prime** over F, if it is not a constant and its divisors are its associates or constants (in F).

For example, $f(x) = x^2-2$ is irreducible over Q (the rational numbers), but is reducible over R (real numbers) since $f(x) = (x-\sqrt{2})(x+\sqrt{2})$.

Definition 7.3
Let f and g be non-zero polynomials. The monic polynomial d, having the properties

(1) d is a divisor of f and g
(2) every polynomial h which is a divisor of f and g, is also a divisor of d

is called the **greatest common divisor** (g.c.d.) of f and g.

We use the notation

$$d = (f,g).$$

Any associate of d is called a *greatest common divisor* of f and g. The above definitions can be generalised to any finite number of polynomials. For example, we write that d is the g.c.d. of the polynomials f, g, and m, as

$$d = (f,g,m).$$

The process used to find the unique g.c.d. of any two polynomials is called the **Euclidean algorithm**.

Theorem 7.2
If $d = (f,g)$, then there exist polynomials u and v such that

$$d(x) = u(x)f(x) + v(x)g(x).$$

Again this result can be generalised to any finite number of polynomials. For example, if $d = (f,g,m)$ then there exist polynomials u, v, w such that

$$d(x) = u(x)f(x) + v(x)g(x) + w(x)m(x).$$

Definition 7.4
The polynomials f and g are said to be **relatively prime** if

$$(f,g) = 1.$$

The definition can be generalised to any finite number of polynomials. For example, the polynomials $f,g,$ and m are relatively prime if

$$(f,g,m) = 1.$$

There are two interesting results which we summarise in the following theorem.

Theorem 7.3
Let s be a polynomial relatively prime to each of the polynomials $f_1,f_2, \ldots f_r$. Let t be an irreducible polynomial which divides the product $f(x)=f_1(x)f_2(x) \ldots f_r(x)$, then

(i) s is relatively prime to the product f, and
(ii) t divides at least one of the polynomials $f_1,f_2, \ldots f_r$.

Theorem 7.4 (The Unique Factorization Theorem)
Every polynomial f can be expressed uniquely as a product of monic prime polynomials $f_1,f_2, \ldots f_k$ in the form

$$f(x) = a[f_1(x)]^{n_1} [f_2(x)]^{n_2} \ldots [f_k(x)]^{n_k}$$

where a is the leading coefficient of f, and n_i $(i=1,2, \ldots k)$ are positive integers.

Theorem 7.5 (The Remainder Theorem)
Let f be a polynomial and α a constant, then

$$f(x) = (x-\alpha)q(x) + r(x),$$

that is, when f is divided by $(x-\alpha)$, the remainder is $r(x)$.

Definition 7.5
f is said to have a **root** or a **zero** $x=\alpha$ if and only if $f(\alpha) = 0$.

7.2 EIGENVALUES OF RATIONAL FUNCTIONS OF A MATRIX

In previous chapters we noted various characteristics associated with a square matrix A such as its eigenvalues, its characteristic equation, its minimum polynomial etc. We shall now investigate how at least some of these characteristics are associated with a polynomial matrix f of A. The first theorem we shall consider proves that a result analogous to Eq (6.5) holds for polynomial functions of a matrix.

Theorem 7.6

If the matrix A of order $n \times n$ has eigenvalues $\lambda_1, \lambda_2, \ldots \lambda_n$, and f is a polynomial of degree r, then

$$|f(A)| = f(\lambda_1)f(\lambda_2)\ldots f(\lambda_n).$$

Proof

We assume that f can be written in terms of factors as

$$f(\lambda) = a(l_1-\lambda)(l_2-\lambda)\ldots(l_r-\lambda)$$

so that

$$f(A) = a[l_1I-A][l_2I-A]\ldots[l_rI-A]. \qquad (7.1)$$

Since each factor $[l_jI-A]$ is a matrix or order $n \times n$, the product of r factors is also a matrix of order $n \times n$, hence

$$|f(A)| = a^n|l_1I-A||l_2I-A|\ldots|l_rI-A|. \qquad (7.2)$$

Since $|\lambda I-A| = C(\lambda)$ (the characteristic equation of A)

$$= (\lambda-\lambda_1)(\lambda-\lambda_2)\ldots(\lambda-\lambda_n)$$

then

$$\begin{aligned}|l_jI-A| &= C(l_j)\\ &= (l_j-\lambda_1)(l_j-\lambda_2)\ldots(l_j-\lambda_n) \qquad (j=1,2,\ldots r)\end{aligned}$$

We can now write (7.2) as

$$\begin{aligned}|f(A)| &= a^n\{[(l_1-\lambda_1)(l_1-\lambda_2)\ldots(l_1-\lambda_n)][(l_2-\lambda_1)(l_2-\lambda_2)\ldots(l_2-\lambda_n)]\\ &\qquad\ldots[(l_r-\lambda_1)(l_r-\lambda_2)\ldots(l_r-\lambda_n)]\}\\ &= \{[a(l_1-\lambda_1)(l_2-\lambda_1)\ldots(l_r-\lambda_1)][a(l_1-\lambda_2)(l_2-\lambda_2)\ldots(l_r-\lambda_2)]\\ &\qquad[a(l_1-\lambda_n)(l_2-\lambda_m)\ldots(l_r-\lambda_n)]\}\\ &= f(\lambda_1)f(\lambda_2)\ldots f(\lambda_n) = \prod_{i=1}^{n} f(\lambda_i)\end{aligned}$$

which proves the theorem.

Example 7.1

Given $A = \begin{bmatrix} 2 & 1 \\ -5 & -2 \end{bmatrix}$ and the polynomial $f(x) = x^2 - 3x + 2$, verify Theorem 7.6.

Solution

In Ex. 6.4 we found that the eigenvalues of A are $\lambda_1 = i$ and $\lambda_2 = -i$, hence

$$f(\lambda_1) = i^2 - 3i + 2 = 1 - 3i$$

$$f(\lambda_2) = i^2 + 3i + 2 = 1 + 3i,$$

and $f(\lambda_1)\,f(\lambda_2) = 10$.

$$\text{Also } f(A) = A^2 - 3A + 2I = \begin{bmatrix} -1 & 0 \\ 0 & -1 \end{bmatrix} - \begin{bmatrix} 6 & 3 \\ -15 & -6 \end{bmatrix} + \begin{bmatrix} 2 & 0 \\ 0 & 2 \end{bmatrix} = \begin{bmatrix} -5 & -3 \\ 15 & 7 \end{bmatrix}$$

so that $|f(A)| = -35 + 45 = 10$.

Actually Theorem 7.6 applies not only when f is a polynomial function but more generally when f is a rational function, a ratio of two polynomials g/h, not necessarily of the same degree but such that $h(A) \neq 0$. Indeed

$$|f(A)| = \frac{|g(A)|}{|h(A)|} = \frac{\prod_{i=1}^{n} g(\lambda_i)}{\prod_{i=1}^{n} h(\lambda_i)} \quad \text{(by application of Theorem 7.6 to both } g \text{ and } h\text{)}$$

$$= \prod_{i=1}^{n} f(\lambda_i).$$

The fact that Theorem 7.6 is applicable not only to polynomial but also to rational functions is not a coincidence. We have from time to time emphasised various differences between manipulations with matrices and scalars; this is another such case. In fact a rational function of a matrix can be reduced to a matrix polynomial form with the aid of the Cayley-Hamilton theorem (or with the aid of any annihilating polynomial).

If $f = g/h$, where g and h are polynomials, then $f(A) = g(A)[h(A)]^{-1}$, where A is a matrix of order $n \times n$ (say).

Let $B = h(A)$.
Since B is a matrix of order $n \times n$ it satisfies its own characteristic equation; that is,

$$B^n + b_1 B^{n-1} + \ldots + b_n I = 0 \quad \text{(if the eigenvalues of } B \text{ are non-zero, } b_n \neq 0)$$

so that $b_n I = -B^n - b_1 B^{n-1} - \ldots - b_{n-1} B$, and

$$[h(A)]^{-1} = B^{-1} = -\frac{1}{b_n}(B^{n-1} + b_1 B^{n-2} + \ldots + b_{n-1} I)$$

$$= \text{a matrix polynomial in } B.$$

It follows that $f(A)$ is the product of two matrix polynomials and hence is itself a matrix polynomial.

Example 7.2

Given $A = \begin{bmatrix} 2 & 1 \\ -5 & -2 \end{bmatrix}$ and $f(x) = \dfrac{x+3}{x-2}$, find the two matrix polynomials whose product is $f(A)$.

Solution

Let $f = g/h$

$$g(A) = A + 3I = \begin{bmatrix} 5 & 1 \\ -5 & 1 \end{bmatrix}$$

and let $B = h(A) = A - 2I = \begin{bmatrix} 0 & 1 \\ -5 & -4 \end{bmatrix}$.

Since $|\lambda I - B| = \begin{bmatrix} \lambda & -1 \\ 5 & \lambda+4 \end{bmatrix} = \lambda^2 + 4\lambda + 5$, then

$$B^2 + 4B + 5I = 0,$$

hence $[h(A)]^{-1} = B^{-1} = -\dfrac{1}{5}[B + 4I] = -\dfrac{1}{5}\begin{bmatrix} 4 & 1 \\ -5 & 0 \end{bmatrix}$.

It follows that $f(A) = \begin{bmatrix} 5 & 1 \\ -5 & 1 \end{bmatrix} \begin{bmatrix} -\frac{4}{5} & -\frac{1}{5} \\ 1 & 0 \end{bmatrix}$.

Example 7.3

With the matrix A of Ex. 7.2 verify the result of Theorem 7.6 when $f(x) = \dfrac{x+3}{x-2}$.

Solution

We have found (see Ex. 6.4 and Ex. 7.1) that the eigenvalues of A are $\lambda_1 = i$ and $\lambda_2 = -i$. Since

$$f(\lambda_1) = \frac{i+3}{i-2} \text{ and } f(\lambda_2) = \frac{-i+3}{-i-2}$$

it follows that

$$f(\lambda_1)\,f(\lambda_2) = \frac{10}{5} = 2.$$

Also from Ex. 7.2

$$f(A) = \begin{bmatrix} -3 & -1 \\ 5 & 1 \end{bmatrix} \text{ so that } |f(A)| = 2,$$

which verifies Theorem 7.6.

In Theorem 6.1 we proved that if A and B are two similar matrices; that is, if

$$B = P^{-1}AP, \text{ then}$$

A and B have the same eigenvalues. We know (see Corollary 1, Theorem 6.1) that if A has eigenvalues $\lambda_1, \lambda_2, \ldots \lambda_n$ then A^r has eigenvalues $\lambda_1^r, \lambda_2^r, \ldots \lambda_n^r$. We can generalise this result and show that if f is a polynomial, then the eigenvalues of $f(A)$ are $f(\lambda_1), f(\lambda_2), \ldots f(\lambda_n)$.

Theorem 7.7

If the matrix A of order $n \times n$ has eigenvalues $\lambda_1, \lambda_2, \ldots \lambda_n$ and f is a polynomial, then the eigenvalues of $f(A)$ are $f(\lambda_1), f(\lambda_2), \ldots f(\lambda_n)$.

Proof

The characteristic equation of the matrix $f(A)$ is

$$|\lambda I - f(A)| = 0.$$

Hence if we let $p(x) = \lambda - f(x)$, the eigenvalues of $f(A)$ are the roots of the equation $|p(A)| = 0$.

Since $p(x)$ is a polynomial, we can apply the result of Theorem 7.6; that is,

$$\begin{aligned} |p(A)| &= p(\lambda_1)\,p(\lambda_2)\ldots p(\lambda_n) \\ &= [\lambda - f(\lambda_1)]\,[\lambda - f(\lambda_2)]\ldots[\lambda - f(\lambda_n)]. \qquad \text{(by (7.2))} \end{aligned}$$

So the roots of $|p(A)| = 0$ are $\lambda = f(\lambda_1)$, $\lambda = f(\lambda_2), \ldots \lambda = f(\lambda_n)$, which proves the theorem.

From the discussion above we note that the theorem holds in the more general case when f is a rational function.

Example 7.4

Given $A = \begin{bmatrix} 2 & 1 \\ -5 & -2 \end{bmatrix}$ and $f(x) = \dfrac{x+3}{x-2}$, find the eigenvalues of $f(A)$ and verify Theorem 7.7.

Solution

In Ex. 7.3, we calculated that

$$f(A) = \begin{bmatrix} -3 & -1 \\ 5 & 1 \end{bmatrix}$$

hence $$|\lambda I - f(A)| = \begin{bmatrix} \lambda+3 & 1 \\ -5 & \lambda-1 \end{bmatrix} = \lambda^2 + 2\lambda + 2 = 0 \text{ for } \lambda = -1 \pm i.$$

Since $\lambda_1 = i$, $\lambda_2 = -i$

$$f(\lambda_1) = \frac{i+3}{i-2} = -1 - i \text{ and } f(\lambda_2) = \frac{-i+3}{-i-2} = -1 + i,$$

which verifies the theorem.

7.3 THE MINIMUM POLYNOMIAL OF A MATRIX

In this section we develop the discussion we began in Sec. 6.6.

Theorem 7.8
The minimum polynomial m for a matrix A divides every annihilating polynomial for A; in particular, m divides the characteristic polynomial C for A.

Proof
Let f be an annihilating polynomial for A. The division algorithm applied to f and m gives

$$f(x) = m(x)q(x) + r(x)$$

where r is either the zero polynomial or $\deg r < \deg m$. As $f(A) = 0$,

$$0 = m(A)q(A) + r(A) = r(A).$$

But unless r is identically zero, the assumption that m is the minimum polynomial for A is contradicted. Hence $f(x) = m(x)q(x)$, and this proves the theorem.

In Ex. 6.13 we found that the characteristic equation for A is

$$C(\lambda) = \lambda^3-2\lambda^2 + \lambda = \lambda(\lambda-1)^2$$

and the minimum polynomial is

$$m(\lambda) = \lambda(\lambda-1).$$

We note that C is divisible by m.

We have taken for granted that the minimum polynomial for a matrix A is unique. Indeed, this is obvious.

Assume that on the contrary that there exist two such polynomials, say m_1 and m_2.

Then $g=m_1-m_2$ is a polynomial of lower degree than m_1 and m_2 (remember m_1, m_2 are monic), and

$$g(A) = m_1(A) - m_2(A) = 0$$

which contradicts the assumption that m is an annihilating polynomial of *lowest* degree. Hence $m_1 = m_2 = m$.

Theorem 7.9

(a) The minimum and the characteristic polynomials of a matrix A have the same distinct linear factors.

(b) If all the eigenvalues of A are distinct, the characteristic and minimum polynomials are the same.

Proof

This proof depends on the fact that the eigenvalues of the zero matrix are zero, and on the result of Theorem 7.7.

(a) Let the characteristic roots of the matrix A of order $n \times n$ be $\lambda_1, \lambda_2, \ldots \lambda_r$ with multiplicities $s_1, s_2, \ldots s_r$ respectively, so that the characteristic equation of A is

$$C(\lambda) = (\lambda-\lambda_1)^{s_1} (\lambda-\lambda_2)^{s_2} \ldots (\lambda-\lambda_r)^{s_r}$$

where $s_i \geq 1$ $(i=1,2,\ldots r)$ and $\sum_{i=1}^{r} s_i = n$. Since the minimum polynomial m for A divides the characteristic polynomial C (Theorem 7.8), it must be of the form

$$m(\lambda) = (\lambda-\lambda_1)^{t_1}(\lambda-\lambda_2)^{t_2} \ldots (\lambda-\lambda_r)^{t_r}$$

where $0 \leq t_i \leq s_i$ $(i=1,2,\ldots r)$. Assume now that $t_j = 0$; that is,

$$m(\lambda_j) \neq 0. \tag{7.3}$$

By Theorem 7.7, the eigenvalues of the matrix $m(A)$ are $m(\lambda_1), m(\lambda_2), \ldots m(\lambda_r)$, and at least one of them $m(\lambda_j) \neq 0$, hence

$$m(A) \neq 0. \tag{7.4}$$

Since (7.4) contradicts the definitions of m, our assumption that $t_j = 0$ is false. Hence $t_j \geq 1$ $(j=1, 2, \ldots r)$, and this proves part (a) of the theorem.

(b) If the eigenvalues of A are all distinct, say $\lambda_1, \lambda_2, \ldots \lambda_n$, then

$$C(\lambda) = (\lambda-\lambda_1) \ \ldots \ (\lambda-\lambda_n),$$

that is, $s_1 = s_2 = \ldots = s_n = 1$.
Since $s_j \geq t_j \geq 1$, it follows that $t_j = 1$ $(j=1, 2, \ldots n)$, and so $C(x) = m(x)$.

Example 7.5

Discuss the results of the above theorem when

(a) $$A = \begin{bmatrix} 1 & 0 & 0 \\ 2 & -1 & 0 \\ 1 & 0 & 2 \end{bmatrix} \quad \text{and} \quad \text{(b) } A = \begin{bmatrix} 1 & 1 & 0 \\ 0 & 0 & 0 \\ 0 & 0 & 0 \end{bmatrix}.$$

Solution

(a) We find $C(\lambda) = |\lambda I - A| = (\lambda-1)(\lambda+1)(\lambda-2)$.
Using the method in Sec. 6.6, we obtain

$$m(\lambda) = \lambda^3 - \lambda^2 - \lambda + 2 = (\lambda-1)(\lambda+1)(\lambda-2).$$

Since each of the linear factors of C are of multiplicity 1, we expect m to have the same factors; this is in fact the case.

(b) $C(\lambda) = |\lambda I - A| = \lambda^2(\lambda-1)$ and $m(\lambda) = \lambda(\lambda-1)$.
Again we note that every factor in m is also a factor in C. On the other hand the multiplicity of the factor corresponding to the zero eigenvalue is $s_1 = 2$, whereas the corresponding multiplicity of this factor in m is $t_1 = 1$.

7.4 DIRECT SUMS AND INVARIANT SUBSPACES

To discuss the case when a matrix A of order $n \times n$ does not have n linearly independent characteristic vectors we make use of the concept of a direct sum of a vector space and that of invariant subspaces.

Definition 7.6
If W_1 and W_2 are subspaces of a vector space V, their *sum* $W_1 + W_2$ is defined as

$$W_1 + W_2 = \{\mathbf{x} = \mathbf{w}_1 + \mathbf{w}_2 \in V; \mathbf{w}_1 \in W_1 \text{ and } \mathbf{w}_2 \in W_2\}.$$

It is not difficult to prove that $W_1 + W_2$ is a subspace of V.

Definition 7.7
If W_1 and W_2 are subspaces of a vector space V, such that

(1) $V = W_1 + W_2$

(2) $W_1 \cap W_2 = \{\mathbf{0}\}$

then we say that V is a **direct sum** of the subspaces W_1 and W_2, and we use the notation

$$V = W_1 \oplus W_2.$$

We also say that W_1 and W_2 are **complementary** and that W_1 is a **complement** of W_2.

Example 7.6
Let $V = R^3$, that is, $V = \{(a, b, c) : a, b, c \in R\}$ and

$$W_1 = \{(a, b, c); a + b + c = 1\}, \qquad W_2 = \{(a, b, 0); a, b \in R\}$$

$$W_3 = \{(a, b, c); a = b = c\}.$$

Show that

(1) $V = W_1 + W_2$, (2) $V = W_1 + W_3$, (3) $V = W_2 + W_3$.

In each case determine whether the sum is direct.

Solution
(1) $R^3 = V = W_1 + W_2$, for if $\mathbf{x} = (a, b, c) \in V$ then we can write

$$\mathbf{x} = (a-d, 1-a-c+d, c) + (d, a+b+c-d-1, 0) \quad \text{(where } d \in R).$$

$$= \mathbf{w}_1 + \mathbf{w}_2 \text{ (say)}$$

where $\mathbf{w}_1 \in W_1$ and $\mathbf{w}_2 \in W_2$.
If $\mathbf{x} \in W_1 \cap W_2$, then $c = 0$ and $a + b = 1$, hence (for example)

$$(\tfrac{1}{2}, \tfrac{1}{2}, 0) \in W_1 \cap W_2.$$

So that although $V = W_1 + W_2$, $V \neq W_1 \oplus W_2$.

(2) $V = W_1 + W_3$, for if $\mathbf{x} = (a, b, c) \in V$ then we can write

$$\mathbf{x} = (a-d, b-d, 1-a-b+2d) + (d, d, d) \quad \left(\text{where } d = \frac{a+b+c-1}{3}\right)$$

that is, $\mathbf{x} = \mathbf{w}_1 + \mathbf{w}_3$ where $\mathbf{w}_1 \in W_1$ and $\mathbf{w}_3 \in W_3$.
If $\mathbf{x} \in W_1 \cap W_3$ then $a = b = c = \tfrac{1}{3}$, so that

$$V = W_1 + W_3, \text{ but } V \neq W_1 \oplus W_3.$$

(3) $V = W_2 + W_3$, for if $\mathbf{x} \in V$ then we can write

$$\mathbf{x} = (a-d, b-e, c) + (d, e, 0) \text{ where } d, e \in R.$$

If $\mathbf{x} \in W_2 \cap W_3$ then $a = b = c$, and $c = 0$; that is, $a = b = c = 0$. Hence $W_2 \cap W_3 = \{\mathbf{O}\}$, so that in this case

$$V = W_2 \oplus W_3.$$

The theorem below makes use of the following well-known result in set theory: If W_1 and W_2 are (finite dimensional) subspaces, then

$$\dim (W_1 + W_2) = \dim W_1 + \dim W_2 - \dim (W_1 \cap W_2).$$

Theorem 7.10

Let W_1 and W_2 be subspaces of a vector space V.

(1) If $V = W_1 \oplus W_2$ then

$$\dim V = \dim W_1 + \dim W_2$$

(it is assumed that W_1, W_2, and V are all finite dimensional).

(2) $V = W \oplus W_2$ if and only if every $\mathbf{x} \in V$ can be written uniquely as

$$\mathbf{x} = \mathbf{w}_1 + \mathbf{w}_2 \text{ where } \mathbf{w}_1 \in W_1 \text{ and } \mathbf{w}_2 \in W_2.$$

Proof

(1) If $V = W_1 \oplus W_2$, then $W_1 \cap W_2 = \{\mathbf{O}\}$, and the result follows.

(2) Assume $V = W_1 \oplus W_2$.
Since $V = W_1 + W_2$ we can write every $\mathbf{x} \in V$ as

$$\mathbf{x} = \mathbf{w}_1 + \mathbf{w}_2 \text{ where } \mathbf{w}_1 \in W_1 \text{ and } \mathbf{w}_2 \in W_2.$$

Also assume that

$$\mathbf{x} = \mathbf{y}_1 + \mathbf{y}_2 \text{ where } \mathbf{y}_1 \in W_1 \text{ and } \mathbf{y}_2 \in W_2, \text{ then}$$

$$\mathbf{w}_1 + \mathbf{w}_2 = \mathbf{y}_1 + \mathbf{y}_2$$

and

$$\mathbf{w}_1 - \mathbf{y}_1 = \mathbf{y}_2 - \mathbf{w}_2.$$

Since

$$\mathbf{z}_1 = \mathbf{w}_1 - \mathbf{y}_1 \in W_1 \text{ and } \mathbf{z}_2 = \mathbf{y}_2 - \mathbf{w}_2 \in W_2$$

and

$$\mathbf{z}_1 = \mathbf{z}_2, \text{ it follows that}$$

$$\mathbf{z}_1, \mathbf{z}_2 \in W_1 \text{ and } W_2, \text{ that is, } \mathbf{z}_1, \mathbf{z}_2 \in W_1 \cap W_2.$$

But

$$\mathbf{W}_1 \cap \mathbf{W}_2 = \{\mathbf{O}\}, \text{ hence } \mathbf{z}_1 = \mathbf{O} = \mathbf{z}_2, \text{ that is,}$$

$$\mathbf{y}_1 = \mathbf{w}_1 \text{ and } \mathbf{y}_2 = \mathbf{w}_2.$$

Conversely, assume that $\mathbf{x} \in V$ and that we can write uniquely

$$\mathbf{x} = \mathbf{w}_1 + \mathbf{w}_2 \text{ where } \mathbf{w}_1 \in W_1 \text{ and } \mathbf{w}_2 \in W_2.$$

Then $V = W_1 + W_2$.
We must show that $W_1 \cap W_2 = \{\mathbf{O}\}$. Suppose that $\mathbf{x} \in W_1 \cap W_2$, then

$$\mathbf{x} = \mathbf{x} + \mathbf{O} \text{ where } \mathbf{x} \in W_1 \text{ and } \mathbf{O} \in W_2$$

and

$$\mathbf{x} = \mathbf{O} + \mathbf{x} \text{ where } \mathbf{x} \in W_2 \text{ and } \mathbf{O} \in W_1.$$

By the uniqueness property it follows that $\mathbf{x} = 0$, that is,

$$V = W_1 \oplus W_2.$$

The definition 7.7 of a direct sum is generalized in the following manner:
If W_1 $W_2, \ldots W_k$ are subspaces of V, then V is a direct sum of the W_i; that is,

$$V = W_1 \oplus W_2 \oplus \ldots \oplus W_k$$

if for each $\mathbf{x} \in V$, we can write

$$\mathbf{x} = \mathbf{w}_1 + \mathbf{w}_2 + \ldots + \mathbf{w}_k \text{ where } \mathbf{w}_i \in W_i \quad (i=1, 2, \ldots k)$$

A closely related concept to the direct sum of vector spaces is that of a direct sum of matrices.

Definition 7.8
The block diagonal matrix A of order $n \times n$

$$A = \begin{bmatrix} A_1 & 0 & 0 & \ldots & 0 \\ 0 & A_2 & 0 & \ldots & 0 \\ 0 & 0 & A_3 & \ldots & 0 \\ \vdots & \vdots & & & \\ 0 & 0 & 0 & \ldots & A_k \end{bmatrix}$$

where A_i is a matrix of order $n_i \times n_i$ (i=1, 2, . . . k), so that $n_1 + n_2 + \ldots + n_k = n$, is called the **direct sum** of the matrices $A_1, A_2, \ldots A_k$ and is denoted by

$$A = A_1 \oplus A_2 \oplus \ldots \oplus A_k.$$

For example, we write

$$A = \begin{bmatrix} 2 & 0 & 0 \\ 0 & 0 & -9 \\ 0 & 1 & 6 \end{bmatrix} \text{ as } [2] \oplus \begin{bmatrix} 0 & -9 \\ 1 & 6 \end{bmatrix}.$$

Example 7.7

Given the matrix $A = A_1 \oplus A_2 \oplus \ldots \oplus A_k$, show that $|A| = |A_1||A_2| \ldots |A_k|$.

Solution

We can obtain the above result by induction. For $k = 1, A = A_1$, hence $|A| = |A_1|$. Assuming the result true for $n = k - 1$, we shall prove it for $n = k$.

We first note the following:

Let B be a square matrix and I_r be the unit matrix of order $r \times r$ such that

$$I_r \oplus B$$

is a matrix of order $n \times n$. On expanding by the first row we have

$$|I_r \oplus B| = 1\ |I_{r-1} \oplus B|.$$

On repeating this process r times, we finally obtain

$$|I_r \oplus B| = |B| = |B \oplus I_r|.$$

Let $B = A_1 \oplus A_2 \oplus \ldots \oplus A_{k-1}$.

Then $A = B \oplus A_k = [B \oplus I_r]\ [I_{n-r} \oplus A_k]$

where A_k is of order $r \times r$. Hence

$$|A| = |B \oplus I_r||I_{n-r} \oplus A_k| = |B|\ |A_k|.$$

The result follows by the hypothesis.

Definition 7.9
Let T be a linear transformations of V into itself. A subspace W of V such that T maps W into itself (that is, when $\mathbf{x} \in W$ then $T(\mathbf{x}) \in W$) is said to be **invariant under** T or T**-invariant**.

In terms of the matrix A representing T (relative to some specified basis), the definition states that:
W is A**-invariant** under the transformation $T : \mathbf{x} \to A\mathbf{x}$ of the vector space V into itself if

$$A\mathbf{x} \in W \text{ whenever } \mathbf{x} \in W.$$

We have already met several examples of invariant subspaces.

If λ_1 is an eigenvalue of A, the corresponding eigenvector $\mathbf{x}_1 \in W$ – the eigenspace of A corresponding to λ_1 (see Def. 6.7). Since $A\mathbf{x}_1 = \lambda_1 \mathbf{x}_1 \in W$, W is A-invariant. Other examples are

(1) $W = \{\mathbf{O}\}$, since whenever $\mathbf{O} \in W$ then $A\mathbf{O} = \mathbf{O} \in W$.
(2) $W = V$ (the vector space itself), since whenever $\mathbf{x} \in V$ then $A\mathbf{x} \in V$.
(3) $W = \ker(T)$, since whenever $\mathbf{x} \in W$, there $A\mathbf{x} = \mathbf{O} \in W$.
(4) $W = \operatorname{Im}(T)$, since if $\mathbf{x} \in W$, it is necessarily true that $A\mathbf{x} \in W$.

Having defined the two concepts, direct sums (of vector spaces and matrices) and A-invariance, we can consider the block-diagonal form of a matrix A (see Def. 7.8) in the following way: A is the matrix representation of a transformation $T : V \to V$. V is a direct sum of A-invariant subspaces $W_1, W_2, \ldots W_k$ having matrix representations $A_1, A_2, \ldots A_k$ (see Theorem 7.11 below).

In our search for a 'simplified' form of a matrix representation of a linear transformation T it seems logical that if we can achieve a block-diagonal form (see Def. 7.8) we will be well on the way to our goal. The following important theorem demonstrates the link between A-invariant spaces and the block diagonal form of A.

Theorem 7.11
Let $T : V \to V$ be a linear transformation having a matrix representation A relative to some specified basis.

The vector space V is a direct sum

$$V = W_1 \oplus W_2 \oplus \ldots \oplus W_k \tag{7.5}$$

of A-invariant W_i of dimension n_i (i=1, 2, . . . k) if and only if there exists a non-singular matrix P such that

$$P^{-1}AP = A_1 \oplus A_2 \oplus \ldots \oplus A_k \tag{7.6}$$

where A_i is a matrix of order $n_i \times n_i$ (i=1, 2, . . . k).

Proof
Assume that (7.5) is true, and that $S_i = \{\mathbf{y}_{i1}, \mathbf{y}_{i2} \ldots \mathbf{y}_{in_i}\}$ is a basis for W_i (i=1, 2, . . . k). Let $S = \{S_1, S_2, \ldots S_k\} = \{\mathbf{y}_{11}, \mathbf{y}_{12}, \ldots \mathbf{y}_{1n_1}, \ldots \mathbf{y}_{k1}, \ldots \mathbf{y}_{k,n_k}\}$. If $\mathbf{x} \in V$, then

$$\mathbf{x} = \mathbf{w}_1 + \mathbf{w}_2 + \ldots + \mathbf{w}_k \text{ where } \mathbf{w}_i \in W_i \quad \text{(see Def. 7.7)}$$

Since S_i is a basis for W_i, we can write the above as

$$\mathbf{x} = (a_{11}\mathbf{y}_{11} + \ldots + a_{1n_1}\mathbf{y}_{1n_1}) + \ldots + (a_{k1}\mathbf{y}_{k1} + \ldots + a_{kn_k}\mathbf{y}_{kn_k}).$$

This shows that S spans V. There is no difficulty in showing that the vectors in S are linearly independent.

It follows that S is a basis for V.

Let P be the matrix whose columns are the vectors $\mathbf{y}_{11}, \mathbf{y}_{12}, \ldots \mathbf{y}_{1n_1}, \ldots \mathbf{y}_{kn_k}$, in that order; that is $P = [\mathbf{y}_{11}, \mathbf{y}_{12}, \ldots \mathbf{y}_{k,n_k}]$. Since $\mathbf{y}_{ij}$ (i=1, 2, . . . k, j=1, 2, . . . n_k) are linearly independent, P is non-singular. As W_1 is A-invariant, and $\mathbf{y}_{11}, \mathbf{y}_{12}, \ldots \mathbf{y}_{1n_{11}} \in W_1$, it follows that

$$A\mathbf{y}_{11}, A\mathbf{y}_{12}, \ldots A\mathbf{y}_{1,n_1} \in W_1,$$

similarly

$$A\mathbf{y}_{21}, A\mathbf{y}_{22}, \ldots A\mathbf{y}_{2,n_2} \in W_2$$

$$\vdots$$

$$A\mathbf{y}_{k,1}, A\mathbf{y}_{k,2}, \ldots A\mathbf{y}_{k,n_k} \in W_k.$$

Since S_1 is a basis for W_1, there exist numbers $a_{1i}, a_{2i}, \ldots a_{n_i,i}$ (i=1, 2, . . . n_1) such that

$$\begin{aligned}
A\mathbf{y}_{11} &= a_{11}\mathbf{y}_{11} + a_{21}\mathbf{y}_{12} + \ldots + a_{n_1,1}\mathbf{y}_{1,n_1} + 0\mathbf{y}_{21} + \ldots + 0.\mathbf{y}_{k,n_k} \\
A\mathbf{y}_{12} &= a_{12}\mathbf{y}_{11} + a_{22}\mathbf{y}_{12} + \ldots + a_{n_1,2}\mathbf{y}_{1,n_1} + 0.\mathbf{y}_{21} + \ldots + 0.\mathbf{y}_{k,n_k} \\
&\vdots \\
A\mathbf{y}_{1,n_1} &= a_{1,n_1}\mathbf{y}_{11} + a_{2,n_1}\mathbf{y}_{12} + \ldots + a_{n_1,n_1}\mathbf{y}_{1,n_1} + \ldots + 0.\mathbf{y}_{1,n_1}.
\end{aligned} \tag{7.7}$$

Since S_2 is a basis for W_2, there exist numbers

$$b_{1i}, b_{2i}, \ldots b_{n_2,i} \ (i=1,2, \ldots n_2) \text{ such that}$$

$$Ay_{21} = 0.y_{11} + \ldots + 0.y_{1,n_1} + b_{11}y_{21} + b_{21}y_{22} + \ldots + b_{n_2,1}y_{2,n_2} + 0.y_{31} + \ldots + 0.y_{k,n_k}$$

$$Ay_{22} = 0.y_{11} + \ldots + 0.y_{1,n_1} + b_{12}y_{21} + b_{22}y_{22} + \ldots + b_{n_2,2}y_{2,n_2} + 0.y_{31} + \ldots + 0.y_{k,n_k}$$

$$\vdots$$

$$Ay_{2,n_2} = 0.y_{11} + \ldots + 0.y_{1,n_1} + b_{1,n_2}y_{21} + b_{2,n_2}y_{22} + \ldots + b_{n_2,n_2}y_{2,n_2} + 0.y_{31} + \ldots + 0.y_{k,n_k}.$$

We continue in a similar manner for $S_2, \ldots S_{k-1}$ and finally note that since S_k is a basis for W_k, there exist numbers

$$t_{1i}, t_{2i}, \ldots t_{n_k,i} \quad (i=1, 2, \ldots n_k) \text{ such that}$$

$$Ay_{k,1} = 0.y_{1,1} + \ldots + 0.y_{k-1,n_{k-1}} + t_{12}y_{k,1} + t_{22}y_{k,2} + \ldots + t_{n_k,2}y_{k,n_k}$$

$$\vdots$$

$$Ay_{k,n_k} = 0.y_{11} + \ldots + 0.y_{k-1,n_{k-1}} + t_{1,n_k}y_{k,1} + t_{2,n_k}y_{k,2} + \ldots + t_{n_k,n_k}y_{k,n_k}.$$

Hence

$$AP = [Ay_{11}, Ay_{12}, \ldots Ay_{k,n_k}]$$

$$= [y_{11}, y_{12}, \ldots y_{k,n_k}] \begin{bmatrix} A_1 & 0 & 0 & \ldots & 0 \\ 0 & A_2 & 0 & \ldots & 0 \\ \vdots & & & & \\ 0 & 0 & 0 & \ldots & A_k \end{bmatrix} \tag{7.8}$$

where

$$A_1 = \begin{bmatrix} a_{11} & a_{12} \ldots & a_{1,n_1} \\ a_{21} & a_{22} & a_{2,n_1} \\ \vdots & \vdots & \vdots \\ a_{n_1,1} & a_{n_1,2} & a_{n_1,n_1} \end{bmatrix}, \qquad A_2 = \begin{bmatrix} b_{11} & b_{12} \ldots & b_{1,n_2} \\ b_{21} & b_{22} & b_{2,n_2} \\ \vdots & \vdots & \vdots \\ b_{n_2,1} & b_{n_2,2} & b_{n_2,n_2} \end{bmatrix},$$

$$\ldots, A_k = \begin{bmatrix} t_{11} & t_{12} \ldots & t_{1,n_k} \\ t_{21} & t_{22} & t_{2,n_k} \\ \vdots & \vdots & \vdots \\ t_{n_k,1} & t_{n_k,2} & t_{n_k,n_k} \end{bmatrix}.$$

It follows from (7.8) that (7.6) holds. Conversely, assume that (7.6) is true and P is non-singular. Then the columns of P, that is, $\mathbf{y}_{11}, \mathbf{y}_{12}, \ldots \mathbf{y}_{k,n_k}$ are linearly independent. The first n_1 of these vectors are a basis for a vector subspace W_1 (say) of V. Writing (7.6) in the form (7.8), on expansion we obtain equations (7.7). These in turn show that W_1 is an A-invariant subspace of V.

Similarly the next n_2 columns of P are a basis for an A-invariant subspace W_2 of V. Continuing in this manner, we conclude that

$$V = W_1 \oplus W_2 \oplus \ldots \oplus W_k.$$

Example 7.8

Given $A = \begin{bmatrix} 14 & -8 & 7 \\ 18 & -11 & 11 \\ 4 & -4 & 5 \end{bmatrix}$, find a matrix P such that $P^{-1}AP$ can be written in a block diagonal form as a direct sum of matrices.

Solution

It seems logical from our previous discussions that we should try to form the matrix P from the eigenvectors of A. The characteristic polynomial of A is

$$C(\lambda) = |\lambda I - A| = \begin{vmatrix} \lambda-14 & 8 & -7 \\ -18 & \lambda+11 & -11 \\ -4 & 4 & \lambda-5 \end{vmatrix} \begin{aligned} &= \lambda^3-8\lambda^2+21\lambda-18 \\ &= (\lambda-2)(\lambda-3)^2 \end{aligned}$$

The characteristic roots are $\lambda_1 = 2$ and $\lambda_2 = \lambda_3 = 3$. Let

$$B_1 = [\lambda_1 I - A] = \begin{bmatrix} -12 & 8 & -7 \\ -18 & 13 & -11 \\ -4 & 4 & -3 \end{bmatrix}.$$

The eigenvectors of A corresponding to $\lambda_1 = 2$, satisfy the equation $B_1\mathbf{y} = \mathbf{O}$. Since $r(B_1) = 2$, [(the first column) = 2 (second column) + 4 (third column)] we expect $(n-r) = 3-2 = 1$ independent solutions (see Theorem 5.1). The solution is $\mathbf{y}_1 = [-1, 2, 4]'$, and $\{\mathbf{y}_1\}$ is the eigenspace corresponding to $\lambda_1 = 2$. Let

$$B_2 = [\lambda_2 I - A] = \begin{bmatrix} -11 & 8 & -7 \\ -18 & 14 & -11 \\ -4 & 4 & -2 \end{bmatrix}.$$

Since $r(B_2) = 2$, [2(first column) = − (second column) + 2(third column)] we expect one independent solution to $B_2\mathbf{y} = \mathbf{O}$, it is $\mathbf{y}_2 = [2, 1, -2]'$, and $\{\mathbf{y}_2\}$ is the eigenspace corresponding to $\lambda_2 = 3$.

In this example we have an eigenvalue $\lambda_2 = 3$ of algebraic multiplicity 2, and the dimension of the corresponding eigenspace is only 1. In this case A is not similar to a diagonal matrix, and V cannot be expressed as a direct sum of three one-dimensional A-invariant subspaces.

Also, we cannot form P from the eigenvectors of A, since we have only two linearly independent ones.

On the other hand, if we let $W_1 = \{\mathbf{y}_1\}$ and $W_2 = \{\mathbf{y}_2, \mathbf{y}_3\}$ where $\mathbf{y}_3 = [-3, 0, 5]'$, then W_1 and W_2 are A-invariant. To check that W_2 is A-invariant we need only note that

$$A\mathbf{y}_3 = [-7, 1, 13]' = \mathbf{y}_2 + 3\mathbf{y}_3 \in W_2 \text{ since } \mathbf{y}_2, \mathbf{y}_3 \in W_2.$$

Hence $V = R^3 = W_1 \oplus W_2$.

Now let $P = [\mathbf{y}_1, \mathbf{y}_2, \mathbf{y}_3] = \begin{bmatrix} -1 & 2 & -3 \\ 2 & 1 & 0 \\ 4 & -2 & 5 \end{bmatrix}$

then $$P^{-1} = \begin{bmatrix} -5 & 4 & -3 \\ 10 & -7 & 6 \\ 8 & -6 & 5 \end{bmatrix}$$ and

$$P^{-1}AP = \begin{bmatrix} -5 & 4 & -3 \\ 10 & -7 & 6 \\ 8 & -6 & 5 \end{bmatrix} \begin{bmatrix} 14 & -8 & 7 \\ 18 & -11 & 11 \\ 4 & -4 & 5 \end{bmatrix} \begin{bmatrix} -1 & 2 & -3 \\ 2 & 1 & 0 \\ 4 & -2 & 5 \end{bmatrix} = \begin{bmatrix} 2 & 0 & 0 \\ 0 & 3 & 1 \\ 0 & 0 & 3 \end{bmatrix}$$

$$= A_1 \oplus A_2$$

where $A_1 = [2]$ and $A_2 = \begin{bmatrix} 3 & 1 \\ 0 & 3 \end{bmatrix}$.

7.5 A DECOMPOSITION OF A VECTOR SPACE

We have seen in the last section that if we can find a decomposition of the vector space V into A-invariant subspaces then we can construct a matrix P such that $P^{-1}AP$ is in a block diagonal form. In this section we shall discuss how such a decomposition of V can be achieved with the aid of the minimal polynomial for A.

Theorem 7.12

Let A be the matrix of $T : V \to V$, a linear transformation of a vector space V over a field F. Let m be the minimal polynomial for A assumed to be the product of two relatively prime monic polynomials f and g, so that

$$m(x) = f(x)g(x).$$

Then $V = W_1 \oplus W_2$

where $W_1 = \ker\,[g(A)]$ and $W_2 = \ker\,[f(A)]$.

Proof

We must show that W_1 (and W_2) are A-invariant subspaces of V, that is, we must show that

$$A\mathbf{x} \in W_i \text{ whenever } \mathbf{x} \in W_i \quad (i = 1, 2).$$

Let $B = f(A)$ so that $W_2 = \ker B$. When $\mathbf{x} \in \ker B$ then $B\mathbf{x} = \mathbf{O}$. Since $BA = f(A).A = A.f(A)$ (see Sec. 6.6), then $B(A\mathbf{x}) = A(B\mathbf{x}) = A\mathbf{O} = \mathbf{O}$, so that $A\mathbf{x} \in \ker B$. It follows that W_2 is A-invariant (and similarly so is W_1).

Since f and g are relatively prime, there exist polynomials p and q (theorem 7.2) such that

$$p(x)f(x) + q(x)g(x) = 1$$

hence

$$p(A)f(A) + q(A)g(A) = I. \tag{7.9}$$

Let $\mathbf{x} \in V$. Multiply (7.9) by $\mathbf{x}$; we obtain

$$\begin{aligned} \mathbf{x} &= p(A)f(A)\mathbf{x} + q(A)g(A)\mathbf{x} \\ &= \mathbf{w}_1 + \mathbf{w}_2 \end{aligned} \tag{7.10}$$

where $\mathbf{w}_1 = p(A)f(A)\mathbf{x}$ and $\mathbf{w}_2 = q(A)g(A)\mathbf{x}$. Now

$$\begin{aligned} g(A)\mathbf{w}_1 &= g(A)p(A)f(A)\mathbf{x} \\ &= f(A)g(A)p(A)\mathbf{x} = m(A)p(A)\mathbf{x} = \mathbf{O} \end{aligned}$$

hence $\mathbf{w}_1 \in W_1 = \ker\,\{g(A)\}$. Similarly

$$f(A)\mathbf{w}_2 = \mathbf{O},$$

hence

$$\mathbf{w}_2 \in W_2 = \ker\,\{f(A)\}.$$

Since $\mathbf{x}$ is an arbitrary element in V, it follows that

$$V = W_1 + W_2. \tag{7.11}$$

To show that the right-hand side of (7.11) is a direct sum, we must show that (Theorem 7.9)

$$\mathbf{x} = \mathbf{w}_1 + \mathbf{w}_2, \quad \mathbf{w}_i \in W_i \quad (i=1, 2)$$

is unique.

Let us assume that we can also express $\mathbf{x}$ as

$$\mathbf{x} = \mathbf{z}_1 + \mathbf{z}_2 \text{ where } \mathbf{z}_i \in W_i \quad (i=1, 2) \tag{7.12}$$

that is,

$$g(A)\mathbf{z}_1 = \mathbf{O} \text{ and } f(A)\,\mathbf{z}_2 = \mathbf{O}, \text{ then}$$

$$\mathbf{O} = g(A)\mathbf{w}_1 - g(A)\mathbf{z}_1 = g(A)\;(\mathbf{w}_1 - \mathbf{z}_1) \text{ and}$$

$$\mathbf{O} = f(A)\mathbf{w}_2 - f(A)\mathbf{z}_2 = f(A)\;(\mathbf{w}_2 - \mathbf{z}_2). \tag{7.13}$$

Also from (7.10) to (7.12) it follows that

$$\mathbf{O} = (\mathbf{w}_1 - \mathbf{z}_1) + (\mathbf{w}_2 - \mathbf{z}_2). \tag{7.14}$$

Multiplying (7.14) on the left by $g(A)$ and $f(A)$ respectively and using (7.13) we obtain for $i = 1, 2$

$$g(A)\,(\mathbf{w}_i - \mathbf{z}_i) = \mathbf{O}$$

and

$$f(A)\,(\mathbf{w}_i - \mathbf{z}_i) = \mathbf{O}.$$

Now multiplying (7.9) by $(\mathbf{w}_i - \mathbf{z}_i)$ we find

$$\mathbf{w}_i - \mathbf{z}_i = p(A)[f(A)\,(\mathbf{w}_i - \mathbf{z}_i)] + q(A)[g(A)\,(\mathbf{w}_i - \mathbf{z}_i)] = \mathbf{O}$$

Thus $\mathbf{w}_i = \mathbf{z}_i$ $(i=1,2)$, so that

$$V = W_1 \oplus W_2 \text{ as required.}$$

We generalise the above result by a repeated application of Theorem 7.12, as in the following corollary.

Corollary

If the minimal polynomial m for the matrix A is a product

$$m(x) = f_1^{r_1}(x)\,f_2^{r_2}(x)\ldots f_k^{r_k}(x)$$

where $f_1(x), f_2(x), \dots f_k(x)$ are irreducible polynomials over F, and $r_1, r_2, \dots r_k$ are positive integers, then

$$V = W_1 \oplus W_2 \oplus \dots \oplus W_k$$

where $W_i = \ker f_i^{r_i}(A)$ $i = 1, 2, \dots k$.

This result is known as the **Primary Decomposition Theorem.**

Example 7.9
With the matrix A of Ex. 7.8, use the Primary Decomposition Theorem to write $V = R^3$ as a direct sum of subspaces W_i.

Solution
We have found that the characteristic polynomial for A is

$$c(x) = (x-2)(x-3)^2.$$

In fact it can be verified (use method of Sec. 6.6) that the minimal polynomial for A is also

$$m(x) = (x-2)(x-3)^2.$$

Hence

$$f_1^{r_1}(x) = x-2, \text{and } f_2^{r_2}(x) = (x-3)^2.$$

Thus $W_1 = \ker\{[A-2I]\}$, and $W_2 = \ker\{[A-3I]^2\}$. We have found (Ex. 7.8) a basis for W_1; it is $\{[-1, 2, 4]'\}$.

To find a basis for W_2, we first calculate

$$[A-3I]^2 = \begin{bmatrix} 11 & -8 & 7 \\ 18 & -14 & 11 \\ 4 & -4 & 2 \end{bmatrix}^2 = \begin{bmatrix} 5 & -4 & 3 \\ -10 & 8 & -6 \\ -20 & 16 & -12 \end{bmatrix} = B^2 \text{ (say)}.$$

Since $r(B^2) = 1$, the number of linearly independent solutions to $B^2\mathbf{x} = \mathbf{O}$ is $3-1 = 2$. Two possible solutions (and hence a basis for W_2) are $\{[2, 1, -2]', [-3, 0, 5]'\}$. We have thus shown that

$$V = R^3 = W_1 \oplus W_2 = \ker\{[A-2I]\} \oplus \ker\{[A-3I]^2\}.$$

The minimal polynomial for a matrix A is seen to be an important concept in the reduction of A into a 'simple' form. In fact if the minimal polynomial is a product of distinct linear factors we can show that A can be reduced into a diagonal form. Before proving this statement we consider a preliminary result which is not strictly necessary to provide an adequate proof but is necessary for the proof given. In any case it is an interesting result in its own right.

Theorem 7.13
Assume that A is a matrix of order $n \times n$ and P is a non-singular matrix such that

$$P^{-1}AP = \Lambda = \text{diag}\{\lambda_1, \lambda_2, \ldots \lambda_n\}.$$

If f is a polynomial of degree m, then

$$f(A) = Pf(\Lambda)P^{-1} \tag{7.15}$$

where

$$f(\Lambda) = \text{diag}\{f(\lambda_1), f(\lambda_2), \ldots f(\lambda_m)\}.$$

Proof
We have already seen (Sec. 7.2) that

$$\Lambda^r = P^{-1}A^rP.$$

This is true for all non-negative integers if we define $\Lambda^0 = \text{diag}\{\lambda_1^0, \lambda_2^0, \ldots \lambda_n^0\} = \text{diag}\{1, 1, \ldots 1\} = I$. Let

$$f(x) = a_0 + a_1x + \ldots + a_mx^m$$

then

$$f(A) = a_0I + a_1A + \ldots + a_mA^m, \quad \text{and}$$

$$P^{-1}f(A)P = a_0P^{-1}P + a_1P^{-1}AP + \ldots + a_mP^{-1}A^mP$$

$$= a_0I + a_1\Lambda + \ldots + a_m\Lambda^m$$

$$= \begin{bmatrix} a_0 & 0 & \ldots & 0 \\ 0 & a_0 & \ldots & 0 \\ \vdots & & & \\ 0 & 0 & \ldots & a_0 \end{bmatrix} + \begin{bmatrix} a_1\lambda_1 & 0 & \ldots & 0 \\ 0 & a_1\lambda_2 & \ldots & 0 \\ \vdots & & & \\ 0 & 0 & \ldots & a_1\lambda_n \end{bmatrix}$$

$$+\ldots+\begin{bmatrix} a_m\lambda_1^m & 0 & \ldots & 0 \\ 0 & a_m\lambda_2^m & \ldots & 0 \\ \vdots & & & \\ 0 & 0 & \ldots & a_m\lambda_n^m \end{bmatrix}$$

$$=\begin{bmatrix} a_0+a_1\lambda_1+\ldots+a_m\lambda_1^m & 0 & \ldots & 0 \\ 0 & a_0+a_1\lambda_2+\ldots+a_m\lambda_2^m & \ldots & 0 \\ \vdots & & & \\ 0 & 0 & \ldots & a_0+a_1\lambda_n+\ldots+a_m\lambda_n^m \end{bmatrix}$$

$$= \text{diag}\,\{f(\lambda_1), f(\lambda_2), \ldots f(\lambda_m)\}.$$

The result (7.15) follows.

Notes

(1) From the above result it is clear that $f(A) = 0$ if

$$f(\lambda_1) = f(\lambda_2) = \ldots = f(\lambda_n) = 0.$$

(2) In the above theorem it is not necessary that all the eigenvalues λ_i of A should be distinct.

Theorem 7.14

The matrix A of order $n \times n$ can be reduced by a similarity transformation to a diagonal form, if and only if the minimal polynomial m for A is a product of distinct linear factors.

Proof

We shall first prove that if A can be reduced to a diagonal form, then m has distinct linear factors. We consider the most general case when the diagonal form has distinct eigenvalues $\lambda_1, \lambda_2, \ldots \lambda_k$ with multiplicities $r_1, r_2 \ldots r_k$ respectively; that is,

$$P^{-1}AP = \text{diag}\{\underbrace{\lambda_1, \lambda_1, \ldots \lambda_1}_{r_1}, \underbrace{\lambda_2, \lambda_2, \ldots \lambda_2}_{r_2}, \ldots \underbrace{\lambda_k, \lambda_k, \ldots \lambda_k}_{r_k}\}.$$

One annihilating polynomial for A is the characteristic polynomial; that is,

$$C(\lambda) = \prod_{i=1}^{k} (\lambda-\lambda_i)^{r_i}.$$

Another annihilating polynomial is

$$f(\lambda) = (\lambda-\lambda_1)(\lambda-\lambda_2)\ldots(\lambda-\lambda_k).$$

To see this we apply the result (7.15) and the fact that by definition of f

$$f(\lambda_1) = f(\lambda_2) = \ldots = f(\lambda_k) = 0.$$

In fact from (7.15) we also note that f is the polynomial of lowest degree which has this property, hence $f = m$, and this proves the first part of the theorem.

Note

It is not necessary to make use of this last statement. It is sufficient to note that f is an annihilating polynomial for A having distinct linear factors. Since m divides every annihilating polynomial for A (Theorem 7.8), m divides f and so the result follows.

Conversely, assume that m is a product of distinct linear factors; that is

$$m(\lambda) = f_1(\lambda)f_2(\lambda)\ldots f_k(\lambda)$$

where

$$f_j(\lambda) = \lambda-\lambda_j \quad (j=1, 2, \ldots k).$$

We know (see the Primary Decomposition Theorem) that

$$V = W_1 \oplus W_2 \oplus \ldots \oplus W_k, \text{ where } W_j = \ker [A-\lambda_j I].$$

If $\mathbf{x} \in W_j$ then $[A-\lambda_j I]\mathbf{x} = \mathbf{0}$ and

$$A\mathbf{x} = \lambda_j \mathbf{x}.$$

That is, $\mathbf{x}$ is an eigenvector of A corresponding to λ_j. We can choose a set S_j of linearly independent vectors

$$S_j = \{\mathbf{x}_{j1}, \mathbf{x}_{j2}, \ldots \mathbf{x}_{jn_j}\}$$

which are a basis for W_j (j=1, 2, . . . k). Each vector in the set is an eigenvector of A. By the process we used in Theorem 7.11 we construct the modal matrix of eigenvectors, which is the union of basis vectors for W_1, W_2, . . . W_k, that is,

$$P = [\mathbf{x}_{11}, \mathbf{x}_{12}, \ldots \mathbf{x}_{k,n_k}].$$

Relative to this basis, we have

$$P^{-1}AP = \Lambda.$$

This proves the theorem.

Example 7.10
Illustrate the result of Theorem 7.14 when

$$A = \begin{bmatrix} -2 & 3 & 0 \\ -4 & 5 & 0 \\ -3 & 3 & 1 \end{bmatrix}.$$

Solution
By the process illustrated in Sec. 6.6 (or otherwise) we can show that the minimal polynomial for A is

$$m(\lambda) = (\lambda-1)(\lambda-2).$$

By Theorem 7.14 the matrix A can be reduced to a diagonal form.

In this example we have (see notation of the Primary Decomposition Theorem)

$$V = W_1 \oplus W_2$$

where

$$W_1 = \ker\{[A-I]\} \text{ and } W_2 = \ker\{[A-2I]\}.$$

Since

$$[A-I] = \begin{bmatrix} -3 & 3 & 0 \\ -4 & 4 & 0 \\ -3 & 3 & 0 \end{bmatrix}, \; W_1 = \{[1, 1, 0]', [0, 0, 1]'\}.$$

Since

$$[A-2I] = \begin{bmatrix} -4 & 3 & 0 \\ -4 & 3 & 0 \\ -3 & 3 & -1 \end{bmatrix}, \; W_2 = \{[3, 4, 3]'\}.$$

Let

$$P = \begin{bmatrix} 1 & 0 & 3 \\ 1 & 0 & 4 \\ 0 & 1 & 3 \end{bmatrix} \text{ then } P^{-1} = \begin{bmatrix} 4 & -3 & 0 \\ 3 & -3 & 1 \\ -1 & 1 & 0 \end{bmatrix}$$

and $P^{-1}AP = \text{diag}\{1, 1, 2\}$.

In one respect it seems that we have now achieved our goal of reducing any matrix A into a 'simple' form by a similarity transformation. We can always transform the matrix into a block-diagonal form, and if the minimal polynomial for A is a product of distinct linear factors, the form is diagonal.

Unfortunately the goal has not yet been reached. The problem is that we have no guarantee that some of the A-invariant subspaces cannot themeselves be split into direct sums of A-invariant subspaces, thus leading to an even 'simpler' block-diagonal representation of A (see Def. 7.8). Even if the vector space V has been split into the maximum number of A-invariant subspaces we do not have (see Ex. 7.9) any particular method for selecting a basis for an A-invariant subspace, and hence for the actual form of the corresponding matrix. We investigate this further in the next section.

7.6 CYCLIC BASES AND THE RATIONAL CANONICAL FORM

In the previous sections we have seen that if the minimal polynomial for a matrix A is in the form

$$m(x) = f_1^{r_1}(x_1)f_2^{r_2}(x_2)\ldots f_k^{r_k}(x_k) \tag{7.16}$$

then the vector space V can be expressed as a direct sum of subspaces W_i. That is,

$$V = W_1 \oplus W_2 \oplus \ldots \oplus W_k$$

where each one of the W_i is related to A by

$$W = \ker\{f^r(A)\} = \ker\{[A - I]^r\}$$

(omitting the suffixes). We form the matrix P from the union of the basis vectors for the W_i and then obtain,

$$P^{-1}AP = A_1 \oplus A_2 \oplus \ldots \oplus A_k.$$

In this section we shall discuss one way of selecting a basis (or indeed sets of bases) for the W_i which will lead to the matrix $P^{-1}AP$ taking a particular form known as the **rational canonical form**. It should be made clear that the particular bases we shall discuss, known as **cyclic bases**, are not the only type possible, nor indeed are the resulting canonical forms the simplest possible. Among various others there are the related bases, known as the **canonical bases** which lead to the matrix $P^{-1}AP$ taking the simpler **Jordan canonical forms** (as in Ex. 7.8). A detailed discussion of other canonical forms can be rather lengthy, but can be found in several of the books mentioned in the Bibliography.

Definition 7.10

Let $\mathbf{y}$ be any (non-zero) vector (of appropriate order) and l the largest integer for which the vectors in the set $S = \{\mathbf{y}, A\mathbf{y}, A^2\mathbf{y}, \ldots A^{l-1}\mathbf{y}\}$ are linearly independent.

The set S is a cyclic basis for an l-dimensional **cyclic space** W.

The vector $\mathbf{y}$ is called a **generator** of W.

The manner of selection of a cyclic basis ensures that the generated space W is A-invariant. To prove this we must show that $A\,\mathbf{x} \in W$, whenever $\mathbf{x} \in W$. Since S is a basis for W, and $\mathbf{x} \in W$, then there exist scalars $a_0, a_1, \ldots a_{l-1}$ such that

$$\mathbf{x} = a_0\mathbf{y} + a_1A\mathbf{y} + \ldots + a_{l-1}A^{l-1}\mathbf{y}$$

and

$$A\mathbf{x} = a_0A\mathbf{y} + a_1A^2\mathbf{y} + \ldots + a_{l-1}A^l\mathbf{y}.$$

If we can show that $A^l\mathbf{y} \in W$, then every term on the right-hand side in the above equation $\in W$ hence $A\mathbf{x} \in W$.

Since by Def. 7.10 the vectors

$$\mathbf{y}, A\mathbf{y}, \ldots A^{l-1}\mathbf{y}, A^l\mathbf{y}$$

are linearly dependent, there exist scalars $b_0, b_1, \ldots b_{l-1}$ (not all zero) such that

$$A^l\mathbf{y} = b_0\mathbf{y} + b_1A\mathbf{y} + \ldots + b_{l-1}A^{l-1}\mathbf{y} \in W. \tag{7.17}$$

It follows that W is A-invariant. If we let

$$f(x) = x^l + c_{l-1}x^{l-1} + \ldots + c_1x + c_0,$$

then from (7.17) with $c_i = -b_i$ $(i=0, 1, \ldots l-1)$ we have

$$f(A)\mathbf{y} = \mathbf{0}.$$

Definition 7.11
A polynomial f having the property that

$$f(A)\mathbf{y} = \mathbf{0}$$

is called an annihilating polynomial of $\mathbf{y}$ with respect to A. The monic polynomial ϕ of lowest degree having the property that

$$\phi(A)\mathbf{y} = \mathbf{0} \tag{7.18}$$

is called the **minimal polynomial** of $\mathbf{y}$ with respect to A.

Since $\mathbf{y}, A\mathbf{y}, \ldots A^{l-1}\mathbf{y}$ are linearly independent, no polynomial of degree less than l can satisfy (7.17), hence the degree of the minimal polynomial of a vector is equal to the dimension of the cyclic space W it generates.

Example 7.11
(1) Find a cyclic basis for the subspace W_2 defined in Ex. 7.8 (and 7.9).
(2) Verify that W_2 is A-invariant.
(3) Determine the minimal polynomial of the generator used.
(4) Is the cyclic basis in (1) unique?

Solution
(1) In Ex. 7.9 the matrix involved was

$$A = \begin{bmatrix} 14 & -8 & 7 \\ 18 & -11 & 11 \\ 4 & -4 & 5 \end{bmatrix} \text{ and } W_2 = \ker\{[A-3I]^2\} = \ker \begin{bmatrix} 5 & -4 & 3 \\ -10 & 8 & -6 \\ -20 & 18 & -12 \end{bmatrix}$$

We found a (non-cyclic) basis for $W_2 = \{[2, 1, -2]', [-3, 0, 5]'\}$. We should attempt to find a cyclic basis by choosing some $\mathbf{y} \in W_2$ and then calculate $A\mathbf{y}$. Let $\mathbf{y} = [1, 2, 1]'$, then $A\mathbf{y} = [5, 7, 1]'$. Both $\mathbf{y}$ and $A\mathbf{y} \in W_2$; they are linearly independent, hence they form a cyclic basis for W_2 (different from our original choice).

(2) Since $\mathbf{y} \in W_2$ and $A\mathbf{y} \in W_2$, it is sufficient to verify that $A(A\mathbf{y}) = \alpha\mathbf{y} + \beta(A\mathbf{y})$ for some scalars α and β. Indeed $A(A\mathbf{y}) = [21, 24, -3]' = -9[1, 2, 1]' + 6[5, 7, 1]'$, and the result follows.

(3) Since $A^2\mathbf{y} = 6A\mathbf{y} - 9\mathbf{y}$

$$\phi(x) = x^2 - 6x + 9 = (x-3)^2.$$

Indeed $A^2\mathbf{y} - 6A\mathbf{y} + 9\mathbf{y} = [A-3I]^2\mathbf{y} = \mathbf{0}$.

(4) The cyclic basis in (1) is not unique. We could choose (among others) the generator $\mathbf{z} = [-1, 1, 3]'$, then $A\mathbf{z} = [-1, 4, 7]'$, and it can be verified that $\{\mathbf{z}, A\mathbf{z}\}$ is a basis for W_2. Since,

$$A^2\mathbf{z} = [3, 15, 15]' = \alpha\mathbf{z} + \beta A\mathbf{z} \text{ for } \alpha = 6 \text{ and } \beta = -9$$

we again have

$$\phi(x) = x^2 - 6x + 9.$$

From the above example it seems that we may be able to find a cyclic basis for any subspace W or a subspace of W associated with a (prime) factor f^r of the minimal polynomial of A.

Indeed this is the case, as we prove in the theorem below. To conform with our previously introduced notation, we shall consider the minimal polynomial of A to be the product

$$m(x) = f^r(x)g(x) \tag{7.19}$$

where f^r is one of the factors of m say $f_i^{r_i}$, where r_i is the largest power of f_i which divides m (see Eq. 7.16) and g is the product of the remaining factors.

It should also be remembered that the factor f in (7.19) is a polynomial of degree 1 over the complex field, but over the real field it could be a polynomial of degree s (say) where $s \geqslant 1$. In what follows we shall assume that f is of degree 1.

Theorem 7.15

If f^r is a factor of the minimal polynomial m for A, then there exists a cyclic subspace U of $W = \ker\{f^r(A)\}$.

Proof
Since $m(x) = f^r(x)g(x)$, it follows that

$$m(A) = 0 = f^r(A)g(A).$$

On the other hand

$f^{r-1}(A)g(A) \neq \mathbf{0}$ (otherwise this would contradict the definition of m).

Hence we can find a vector $\mathbf{x}$ such that

$$[f^{r-1}(A)g(A)]\,\mathbf{x} \neq \mathbf{0} \tag{7.20}$$

Let $\mathbf{y} = g(A)\mathbf{x}$.
Then $f^r(A)\mathbf{y} = f^r(A)g(A)\mathbf{x} = m(A)\mathbf{x} = \mathbf{0}$, hence $\mathbf{y} \in \ker\{f^r(A)\}$, but

$$f^{r-1}(A)\mathbf{y} = f^{r-1}(A)g(A)\mathbf{x} \neq \mathbf{0} \quad \text{(by (7.20))}.$$

It follows that (see Def. 7.11) f^r is the minimum polynomial of $\mathbf{y}$. If

$$f^r(x) = (x-\lambda)^r = x^r + a_{r-1}x^{r-1} + \ldots + a_0$$

= polynomial of degree r, then the space U associated with f^r has the cyclic basis $\{\mathbf{y}, A\mathbf{y}, \ldots A^{r-1}\mathbf{y}\}$ and has dimension r. There are two possibilities:

(1) $U = W$ as in Ex. 7.11, where we found a cyclic basis for $W = \ker\{[A-3I]^2\}$. More generally this is the case whenever

$$\dim U = r = \dim \ker \{f^r(A)\}.$$

(2) $U < W$; this is the case when

$$\dim U = r < \dim \ker \{f^r(A)\}.$$

This second possibility is illustrated by the following example:

Example 7.12
Determine the minimal polynomial for the matrix

$$A = \begin{bmatrix} 1 & 0 & 1 \\ -1 & 2 & 1 \\ -1 & 0 & 3 \end{bmatrix}.$$

Illustrate that in this example the case (2) of Theorem 7.15 applies.

Solution
It is not difficult to verify that the minimal polynomial for A is

$$m(x) = (x-2)^2$$

(the characteristic polynomial is $c(x) = (x-2)^3$). Using the terminology of Theorem 7.14, we have

$$f^r(x) = (x-2)^2, \text{ that is, } r = 2,$$

therefore $\dim U = 2$.

On the other hand

$$W = \ker\{f^r(A)\} = \ker\{m(A)\} = \ker\{0\}$$

so that $\dim W = 3$.

It follows that in this case we can find a cyclic basis for a subspace U of W but not for W itself. To illustrate this, we choose $\mathbf{y} = [1, 0, 0]'$ then $A\mathbf{y} = [1, -1, -1]'$ and $A^2\mathbf{y} = [0, -4, -4]$. Since $A^2\mathbf{y} = -4\mathbf{y} + 4A\mathbf{y}$, the set $\{\mathbf{y}, A\mathbf{y}, A^2\mathbf{y}\}$ is not a basis for W.

Although in general, that is, when $\dim U < \dim \ker\{f^r(A)\}$ we cannot find a cyclic basis for W, we can express W as a *direct sum of cyclic subspaces* which are irreducible, that is, cannot be further decomposed into direct sums of A-invariant subspaces. The following theorem, which we do not prove, indicates what is involved.

Theorem 7.16
Let f^r be a factor of the minimal polynomial for A (see Theorem 7.15 for notation). There exist cyclic subspaces $U_1, U_2, \ldots U_j$ of dimensions $r_1, r_2, \ldots r_j$ respectively where $r = r_1 \geqslant r_2 \geqslant \ldots \geqslant r_j$ such that

$$W = \ker\{f^r(A)\} = U_1 \oplus U_2 \oplus \ldots \oplus U_j. \tag{7.21}$$

Briefly, the theorem states the following: We know (Theorem 7.15) that there exists a cyclic subspace U_1 of W whose generator is a vector $\mathbf{y}_1$ (say). Assume that

$$\dim U_1 = r = r_1.$$

If $\dim W = r_1$, then $U_1 = W$, the process is ended, and W is a cyclic space. If $\dim W > r_1$, it can be shown that there exists a generator $\mathbf{y}_2 \in W$ of a subspace U_2 and an integer r_2 such that

$$\mathbf{y}_2 \notin U_1, f^{r_2}(A)\,\mathbf{y}_2 = \mathbf{O} \in U_1 \text{ and } f^{r_2-1}(A)\mathbf{y}_2 \in U_1$$

and
(1) $r_1 \geq r_2 \geq 1$
(2) $U_1 + U_2$ is a direct sum.
If $U_1 \oplus U_2 = W$, the process is ended. Otherwise we continue until we can express W as a direct sum of subspaces $U_1, U_2, \ldots U_j$.

Example 7.13
For the matrix of Ex. 7.12, express $W = \ker \{f^r(A)\}$ as a direct sum of cyclic spaces U_1 and U_2.

Solution
We have found a cyclic basis for U_1; it is $\{[1, 0, 0]', [1, -1, -1]'\}$. To find a cyclic basis for U_2, we must find a vector $\mathbf{y}_2$ such that

$$\mathbf{y}_2 \notin U_1 \text{ and } f^{r_2}(A)\mathbf{y}_2 = \mathbf{O} \in U_1$$

where $r_2 = 1$ (as $r_1 = 2$ and $r_1 + r_2 = 3 = \dim W$).

Since $f(A) = [A-2I] = \begin{bmatrix} -1 & 0 & 1 \\ -1 & 0 & 1 \\ -1 & 0 & 1 \end{bmatrix}$

and we choose $\mathbf{y}_2 = [0, 1, 0]'$, then both conditions are satisfied and $\{\mathbf{y}_2\}$ is a cyclic basis for U_2.

Although in the above example there was no difficulty in choosing $\mathbf{y}_2$, in general the situation can be more complicated since $\mathbf{y}_2$ must satisfy two conditions. There exist methods which help in the search for an appropriate vector, but we shall not discuss them here.

We shall now review briefly how near we have reached out goal of selecting an appropriate basis relative to which the transformation T is represented by a 'simple' matrix. In Theorem 7.11 we proved that if we can express the vector space V as a direct sum of A-invariant subspaces $W_1, W_2, \ldots W_k$ (Eq. (7.5)), then the matrix A can be expressed in a 'simple', that is, a block diagonal form (Eq. (7.6)). The Primary Decomposition Theorem allows us to express V as a direct sum of subspaces, but it gives no indication of how to select bases for the W_i (i=1, 2, . . . k) and hence leaves the form of the diagonal elements $A_1, A_2, \ldots A_k$ of the block diagonal matrix undefined.

The choice of cyclic bases ensures that each W_i can be expressed as a direct sum of irreducible cyclic spaces $U_1, U_2, \ldots U_j$. Although the bases for the U_i are not unique (since they depend on the choice of the generators) the use of cyclic bases does define the form of the matrices $A_1, A_2, \ldots A_k$. Theoretically, the problem appears extremely complicated, since the transformation matrix P, such that

$$P^{-1}AP = A_1 \oplus A_2 \oplus \ldots \oplus A_k \tag{7.22}$$

is formed as

$$P = [S_1, S_2, \ldots S_k]$$

where S_i is the set of basis vectors for W_i. Since $W_i = U_{i1} \oplus U_{i2} \oplus \ldots \oplus U_{ij}$, then

$$S_i = \{Q_{i1}, Q_{i2}, \ldots Q_{ij}\}$$

where Q_{is} is the cyclic basis for U_{is} (s=1, 2, . . . j), and each A_i in (7.22) has the form

$$A_i = B_{i1} \oplus B_{i2} \oplus \ldots \oplus B_{ij} \tag{7.23}$$

(actually the suffix j can be a different number for each W). Practically, the situation is of course more tractable.

Having said that the cyclic basis does define the form of the matrix A_i (i=1, 2, . . . k) we shall now investigate what this form is. For simplicity we shall drop the suffix (i) from the space U_i and from the corresponding matrix representation B_i.

We assume that the cyclic subspace U of $W = \ker\{f^r(A)\}$ has dimension l and generator $\mathbf{y}$, that is, its basis is

$$\{\mathbf{y}, A\mathbf{y}, A^2\mathbf{y}, \ldots A^{l-1}\mathbf{y}\}. \tag{7.24}$$

Let the minimal polynomial of $\mathbf{y}$ with respect to A be

$$\phi(x) = x^l + a_{l-1}x^{l-1} + \ldots + a_1x + a_0,$$

that is,

$$\phi(A)\mathbf{y} = A^l\mathbf{y} + a_{l-1}A^{l-1}\mathbf{y} + \ldots + a_1A\mathbf{y} + a_0\mathbf{y} = \mathbf{0}. \tag{7.25}$$

Since $P^{-1}AP = A_1 \oplus \ldots \oplus A_k$,

$$AP = P[A_1 \oplus \ldots \oplus A_k]. \tag{7.26}$$

We are concentrating on the part-basis of P which is the cyclic basis for U and which has the matrix representation B relative to this basis. For this 'restricted' matrix, P, (7.26) becomes

$$A[\mathbf{y}, A\mathbf{y}, \ldots A^{l-1}\mathbf{y}] = [\mathbf{y}, A\mathbf{y}, \ldots A^{l-1}\mathbf{y}]B. \tag{7.27}$$

But

$$A.\mathbf{y} = 0.\mathbf{y} + 1.A\mathbf{y} + 0.A^2\mathbf{y} + \ldots + 0.A^{l-1}\mathbf{y}$$

$$A.(A\mathbf{y}) = 0.\mathbf{y} + 0.A\mathbf{y} + 1.A^2\mathbf{y} + \ldots + 0.A^{l-1}\mathbf{y}$$

$$\vdots$$

$$A.(A^{l-2}\mathbf{y}) = 0.\mathbf{y} + 0.A\mathbf{y} + 0.A^2\mathbf{y} + \ldots + 1.A^{l-1}\mathbf{y}$$

and from (7.25)

$$A.(A^{l-1}\mathbf{y}) = -a_0\mathbf{y} - a_1A\mathbf{y} - a_2A^2\mathbf{y} - \ldots - a_{l-1}A^{l-1}\mathbf{y}$$

so that

$$A[\mathbf{y}, A\mathbf{y}, A^2\mathbf{y}, \ldots A^{l-1}\mathbf{y}] = [\mathbf{y}, A\mathbf{y}, A^2\mathbf{y}, \ldots A^{l-1}\mathbf{y}]\begin{bmatrix} 0 & 0 & \ldots & 0 & -a_0 \\ 1 & 0 & & 0 & -a_1 \\ 0 & 1 & & 0 & -a_2 \\ \vdots & & & \vdots & \vdots \\ 0 & 0 & & 1 & -a_{l-1} \end{bmatrix},$$

that is, the form of the matrix B is

$$B = \begin{bmatrix} 0 & 0 & \ldots & 0 & -a_0 \\ 1 & 0 & & 0 & -a_1 \\ 0 & 1 & & 0 & -a_2 \\ \vdots & & & & \\ 0 & 0 & & 1 & -a_{l-1} \end{bmatrix} \tag{7.28}$$

where $a_0, a_1, \ldots a_{l-1}$ are the coefficients of the minimal polynomial of $\mathbf{y}$ (with respect to A).

Definition 7.12
The matrix B is called the **companion matrix** of the polynomial

$$\phi(x) = x^l + a_{l-1}x^{l-1} + \ldots + a_1 x + a_0.$$

Definition 7.13
A matrix A is said to be in a **rational canonical form** if it is in a block diagonal form $A_1 \oplus A_2 \oplus \ldots \oplus A_k$ where each A_i $(i=1,2,\ldots k)$ is in the form of a companion matrix.

Example 7.14
Find the rational canonical form of the matrix $A = \begin{bmatrix} 1 & 0 & 1 \\ -1 & 2 & 1 \\ -1 & 0 & 3 \end{bmatrix}$ (see Ex. 7.12 and 7.13).

Solution
In Ex. 7.12 we have shown that the minimal polynomial for A is

$$m(x) = (x-2)^2,$$

hence

$$V = R^3 = W = \ker\{[A-2I]^2\}.$$

In Ex. 7.13, we found that

$$W = U_1 \oplus U_2$$

where $\{[1, 0, 0]', [1, -1, -1]'\}$ and $\{[0, 1, 0]'\}$ are the bases for U_1 and U_2 respectively. Since the minimal polynomial of $\mathbf{y}_1$ (the generator of U_1) is

$$(x-2)^2 = x^2 - 4x + 2$$

the corresponding companion matrix (see Eq. (7.28)) is

$$\begin{bmatrix} 0 & -2 \\ 1 & 4 \end{bmatrix}.$$

Since the minimal polynomial of $\mathbf{y}_2$ (the generator of U_2) is

$$x - 2$$

the corresponding companion matrix is

$$[2].$$

By the above discussion we should have

$$P^{-1}AP = B_1 \oplus B_2 = \begin{bmatrix} 0 & -2 \\ 1 & 4 \end{bmatrix} \oplus [2] = \begin{bmatrix} 0 & -2 & 0 \\ 1 & 4 & 0 \\ 0 & 0 & 2 \end{bmatrix}. \quad (7.29)$$

P is the matrix of the cyclic basis vectors for U_1 and U_2, that is,

$$P = \begin{bmatrix} 1 & 1 & 0 \\ 0 & -1 & 1 \\ 0 & -1 & 0 \end{bmatrix}.$$

(Notice that the 'restricted' matrix P referred to in the above discussion is the matrix of the cyclic basis vectors for U_1, that is,

$$\begin{bmatrix} 1 & 1 \\ 0 & -1 \\ 0 & -1 \end{bmatrix}).$$

It is not difficult to verify that

$$P^{-1} = \begin{bmatrix} 1 & 0 & 1 \\ 0 & 0 & -1 \\ 0 & 1 & -1 \end{bmatrix}$$

and that $P^{-1}AP$ is the matrix (7.29).

Example 7.15
Find the minimal polynomial for the matrix

$$A = \begin{bmatrix} 14 & -8 & 7 \\ 18 & -11 & 11 \\ 4 & -4 & 5 \end{bmatrix}$$

and express $V=R^3$ as a direct sum of cyclic spaces.

Hence find a matrix P such that $P^{-1}AP$ is in the rational canonical form. Verify the result.

Solution
In Ex. 7.9 it was shown that the minimal molynomial for A is

$$m(x) = (x-2)(x-3)^2,$$

hence

$$V = R^3 = W_1 \oplus W_2$$

where $W_1 = \ker\{[A-2I]\}$ and $W_2 = \ker\{[A-3I]^2\}$. We also found (Ex. 7.8) that

$$W_1 = \{\mathbf{y}_1 = [-1, 2, 4]'\}.$$

Since $\dim W_1 = 1$, $\mathbf{y}_1$ is a generator of the cyclic space W_1. Also (Ex. 7.9), we found a cyclic space for W_2. $W_2 = \{\mathbf{y}_2 = [1, 2, 1]', A\mathbf{y}_2 = [5, 7, 1]'\}$. Since the minimal polynomial of $\mathbf{y}_1$ is

$$x-2$$

the corresponding companion matrix (7.28) is

$$A_1 = [2].$$

Similarly the minimal polynomial of $\mathbf{y}_2$ is

$$(x-3)^2 = x^2 - 6x + 9$$

and the corresponding companion matrix is

$$A_2 = \begin{bmatrix} 0 & -9 \\ 1 & 6 \end{bmatrix}.$$

It follows that

$$P^{-1}AP = A_1 \oplus A_2 = [2] \oplus \begin{bmatrix} 0 & -9 \\ 1 & 6 \end{bmatrix} = \begin{bmatrix} 2 & 0 & 0 \\ 0 & 0 & -9 \\ 0 & 1 & 6 \end{bmatrix}. \tag{7.30}$$

To check this result, we form the matrix P from the cyclic basis vectors for W_1 and W_2, that is,

$$P = \begin{bmatrix} -1 & 1 & 5 \\ 2 & 2 & 7 \\ 4 & 1 & 1 \end{bmatrix} \text{ so that } P^{-1} = \begin{bmatrix} -5 & 4 & -3 \\ 26 & -21 & 17 \\ -6 & 5 & -4 \end{bmatrix}.$$

On multiplying out, we find that $P^{-1}AP$ in the matrix (7.30).

Notice that if in the above example we construct P from the cyclic basis vectors of W_2 followed by the one for W_1, that is,

$$P = \begin{bmatrix} 1 & 5 & -1 \\ 2 & 7 & 2 \\ 1 & 1 & 4 \end{bmatrix} \text{ so that } P^{-1} = \begin{bmatrix} 26 & -21 & 17 \\ -6 & 5 & -4 \\ -5 & 4 & -3 \end{bmatrix}$$

we then obtain

$$P^{-1}AP = \begin{bmatrix} 0 & -9 \\ 1 & 6 \end{bmatrix} \oplus [2] = \begin{bmatrix} 0 & -9 & 0 \\ 1 & 6 & 0 \\ 0 & 0 & 2 \end{bmatrix}.$$

7.7 THE JORDAN CANONICAL FORMS

We have seen in the previous sections that relative to a cyclic basis for a cyclic space U, the corresponding matrix has the companion form (7.28). On the other hand we can find bases, other than cyclic, for U.

Consider a cyclic subspace U of $W = \ker \{f^r(A)\}$ of dimension l. By Theorem 7.14 we know that the minimal polynomial ψ of $\mathbf{y} \in W$, has the form

$$\psi(x) = f^l(x) \text{ where } l \leq r \quad (\text{depending on whether } U = W \text{ or } U \subset W)$$

so that $f^l(A)\mathbf{y} = \mathbf{0}$. (7.31)

Definition 7.14
The set of vectors $C = \{\mathbf{y}, f(A)\mathbf{y}, \ldots f^{l-1}(A)\mathbf{y}\}$ forms a basis, called the **canonical basis** for the l-dimensional cyclic subspace U of $W = \ker \{f^r(A)\}$, with generator $\mathbf{y}$.

To show that the set C is a basis for U, we would need to show that the l vectors in C are
(1) vectors $\in U$
(2) linearly independent.

(1) We know that the cyclic basis for U is $S = \{\mathbf{y}, A\mathbf{y}, \ldots A^{l-1}\mathbf{y}\}$. By the assumption that the field of scalars is complex, we have

$$f(x) = x - \lambda.$$

Hence for $s = 1, 2, \ldots l-1$,

$$f^s(x) = \text{polynomial of degree } s \text{ in } x.$$

It follows that $f^s(A)\mathbf{y}$ is a vector which can be written as a linear combination of $\mathbf{y}, A\mathbf{y}, \ldots A^s\mathbf{y}$, hence

$$f^s(A)\mathbf{y} \in U \text{ for } s = 0, 1, \ldots l-1.$$

(2) Assume that for some $s \leq l$, there exist scalars $b_0, b_1, \ldots b_{s-1}$ (not all zero) such that

$$f^s(A)\mathbf{y} = b_{s-1}f^{s-1}(A)\mathbf{y} + \ldots + b_1 f(A)\mathbf{y} + b_0\mathbf{y}$$

that is

$$f^s(A)\mathbf{y} - b_{s-1}f^{s-1}(A)\mathbf{y} \pm \ldots - b_0\mathbf{y} = \mathbf{0}.$$

Hence $g(A)\mathbf{y} = \mathbf{0}$,

where g is a polynomial of degree s. This implies that g is an annihilating polynomial of $\mathbf{y}$ (Def. 7.11) of *lower* degree than ψ the minimal polynomial of $\mathbf{y}$. This is impossible, so that our assumption is false.

It follows that the set C is a basis for U.

We now discuss (see page 211) the resulting form of the matrix $A_i (i = 1, 2, \ldots k)$ relative to the basis C (we drop the suffix i from A_i).

We use the relations $f(x) = x - \lambda$, so that $f(A) = A - \lambda I$ and $A = \lambda I + f(A)$.

$$
\begin{aligned}
\text{Thus } A\mathbf{y} &= [\lambda I + f(A)]\,\mathbf{y} &&= \lambda\mathbf{y} + f(A)\mathbf{y} \\
Af(A)\mathbf{y} &= [\lambda I + f(A)]f(A)\mathbf{y} &&= \lambda f(A)\mathbf{y} + f^2(A)\mathbf{y} \\
&\vdots \\
Af^{l-1}(A)\mathbf{y} &= [\lambda I + f(A)]f^{l-1}(A)\mathbf{y} &&= \lambda f^{l-1}(A)\mathbf{y} + f^l(A)\mathbf{y} \\
& &&= \lambda f^{l-1}(A)\mathbf{y} \quad (\text{by } (7.31)).
\end{aligned}
$$

Hence

$$
A[\mathbf{y}, f(A)\mathbf{y}, \ldots f^{l-1}(A)\mathbf{y}] = [\mathbf{y}, f(A)\mathbf{y}, \ldots f^{l-1}(A)\mathbf{y}]
\begin{bmatrix}
\lambda & 0 & \cdots & 0 & 0 \\
1 & \lambda & & 0 & 0 \\
0 & 1 & & 0 & 0 \\
\vdots & & & & \\
0 & 0 & & \lambda & 0 \\
0 & 0 & & 1 & \lambda
\end{bmatrix},
$$

that is, corresponding to the matrix B of Eq. (7.28),

$$
J = \begin{bmatrix}
\lambda & 0 & \cdots & 0 & 0 \\
1 & \lambda & & 0 & 0 \\
0 & 1 & & 0 & 0 \\
\vdots & \vdots & & \vdots & \vdots \\
0 & 0 & & \lambda & 0 \\
0 & 0 & & 1 & \lambda
\end{bmatrix}. \tag{7.32}
$$

Definition 7.15
The matrix J in (7.32) is said to be in a **Jordan canonical form.**

Definition 7.16
A matrix A is said to be in a **Jordan block form** if it is in a block diagonal form $A_1 \oplus A_2 \oplus \ldots \oplus A_k$ where each A_i $(i = 1, 2, \ldots k)$ is in a Jordan canonical form.

Example 7.16
Find the Jordan block form of the matrix

$$A = \begin{bmatrix} 14 & -8 & 7 \\ 18 & -11 & 11 \\ 4 & -4 & 5 \end{bmatrix}. \qquad \text{(see Ex. 7.15)}$$

Verify the result.

Solution
Following the solution for Ex. 7.15, we have

$$m(x) = (x-2)(x-3)^2$$

$$V = W_1 \oplus W_2$$

where $W_1 = \ker\{[A-2I]\}$ and $W_2 = \ker\{[A-3I]^2\}$.

Corresponding to the factor $x - 2$ the Jordan form is

$$A_1 = [2].$$

Corresponding to the factor $(x-3)^2$, the Jordan form is

$$A_2 = \begin{bmatrix} 3 & 0 \\ 1 & 3 \end{bmatrix}.$$

Hence we can transform A to the Jordan block

$$A_1 \oplus A_2 = [2] \oplus \begin{bmatrix} 3 & 0 \\ 1 & 3 \end{bmatrix} = \begin{bmatrix} 2 & 0 & 0 \\ 0 & 3 & 0 \\ 0 & 1 & 3 \end{bmatrix}. \qquad (7.33)$$

As in Ex. 7.15, we can choose

$$W_1 = \{\mathbf{y}_1 = [-1, 2, 4]'\}.$$

For a canonical basis for W_2 we can choose $\{\mathbf{y}_2, f(A)\mathbf{y}_2\} = \{[1, 2, 1]', [2, 1, -2]'\}$ where

$$f(A) = [A-3I] = \begin{bmatrix} 11 & -8 & 7 \\ 18 & -14 & 11 \\ 4 & -4 & 2 \end{bmatrix}.$$

To verify the result, we form the matrix of the basis vectors

$$P = \begin{bmatrix} -1 & 1 & 2 \\ 2 & 2 & 1 \\ 4 & 1 & -2 \end{bmatrix} \text{ and calculate } P^{-1} = \begin{bmatrix} -5 & 4 & -3 \\ 8 & -6 & 5 \\ -6 & 5 & -4 \end{bmatrix}.$$

We then find that

$$P^{-1}AP \text{ is the matrix (7.33).}$$

Note 1

It is sometimes useful to note that a matrix in Jordan form is the sum of two matrices of a rather simple tructure, for example

$$\begin{bmatrix} \lambda & 0 & 0 \\ 1 & \lambda & 0 \\ 0 & 1 & \lambda \end{bmatrix} = \lambda \begin{bmatrix} 1 & 0 & 0 \\ 0 & 1 & 0 \\ 0 & 0 & 1 \end{bmatrix} + \begin{bmatrix} 0 & 0 & 0 \\ 1 & 0 & 0 \\ 0 & 1 & 0 \end{bmatrix} = \lambda I + L$$

where L has unit elements on the first infradiagonal, (that is, the diagonal below the main diagonal) and zeroes elsewhere.

Note 2

To be precise, the matrix of Eq. (7.32) is said to be in a **lower Jordan canonical** form.

On the other hand, if we choose the basis for the cyclic space in the reversed order, that is,

$$C' = \{f^{l-1}(A)\mathbf{y}, f^{l-2}(A)\mathbf{y}, \ldots f(A)\mathbf{y}, \mathbf{y}\}$$

then the matrix takes the form

$$J = \begin{bmatrix} \lambda & 1 & \cdots & 0 & 0 \\ 0 & \lambda & & 0 & 0 \\ 0 & 0 & & 0 & 0 \\ \vdots & \vdots & & \vdots & \vdots \\ 0 & 0 & & 1 & 0 \\ 0 & 0 & & \lambda & 1 \\ 0 & 0 & & 0 & \lambda \end{bmatrix}. \tag{7.34}$$

In (7.34) the matrix is said to be in the **upper Jordan canonical form**. We can write a matrix in an upper Jordan canonical form as a sum of two simpler matrices, for example

$$\begin{bmatrix} \lambda & 1 & 0 \\ 0 & \lambda & 1 \\ 0 & 0 & \lambda \end{bmatrix} = \lambda \begin{bmatrix} 1 & 0 & 0 \\ 0 & 1 & 0 \\ 0 & 0 & 1 \end{bmatrix} + \begin{bmatrix} 0 & 1 & 0 \\ 0 & 0 & 1 \\ 0 & 0 & 0 \end{bmatrix} = \lambda I + U$$

where U has unit elements along the first super diagonal, and zeroes elsewhere.

If in Ex. 7.16, we reverse the order of the basis vectors for W_2, the corresponding matrix is

$$P = \begin{bmatrix} -1 & 2 & 1 \\ 2 & 1 & 2 \\ 4 & -2 & 1 \end{bmatrix} \text{ so that } P^{-1} = \begin{bmatrix} -5 & 4 & -3 \\ -6 & 5 & -4 \\ 8 & -6 & 5 \end{bmatrix}$$

and

$$P^{-1}AP = \begin{bmatrix} 2 & 0 & 0 \\ 0 & 3 & 1 \\ 0 & 0 & 3 \end{bmatrix} = [2] \oplus \begin{bmatrix} 3 & 1 \\ 0 & 3 \end{bmatrix}$$

where the second matrix in the above equation is in the upper Jordan form.

Example 7.17

Find the matrix P, such that $P^{-1}AP$ is in the (upper) Jordan Block form, where

$$A = \begin{bmatrix} 1 & 0 & 1 \\ -1 & 2 & 1 \\ -1 & 0 & 3 \end{bmatrix}. \qquad \text{(see Exs. 7.12, 7.13, and 7.14).}$$

Solution

In Exs. 7.12 and 7.13 we have found $W = \ker \{f^2(A)\} = U_1 \oplus U_2$ where U_1 and U_2 are cyclic spaces. The canonical basis for U_1 is $\{f(A)\mathbf{y}_1, \mathbf{y}_1\}$ where

$$f(A) = [A - 2I] = \begin{bmatrix} -1 & 0 & 1 \\ -1 & 0 & 1 \\ -1 & 0 & 1 \end{bmatrix}, \text{ so that}$$

$U_1 = \{[-1, -1, -1]', [1, 0, 0]'\}$.

We have found a basis for U_2; it is $\{[0, 1, 0]'\}$.

Hence

$$P = \begin{bmatrix} -1 & 1 & 0 \\ -1 & 0 & 1 \\ -1 & 0 & 0 \end{bmatrix} \text{ so that } P^{-1} = \begin{bmatrix} 0 & 0 & -1 \\ 1 & 0 & -1 \\ 0 & 1 & -1 \end{bmatrix}$$

and

$$P^{-1}AP = \begin{bmatrix} 2 & 1 & 0 \\ 0 & 2 & 0 \\ 0 & 0 & 2 \end{bmatrix} = \begin{bmatrix} 2 & 1 \\ 0 & 2 \end{bmatrix} \oplus [2].$$

7.8 MATRIX FUNCTIONS

We have already considered (see Sections 6.8 and 7.2) polynomial and rational functions of a square matrix A defined by analogy with corresponding scalar functions. We shall now take this process a stage further and define other matrix functions.

It is well known that if z is a complex number the power series

$$\sum_{n=0}^{\infty} a_n z^n$$

converges to a function $f(z)$ for $|z| < R$ where $R = \lim_{n\to\infty} \dfrac{a_n}{a_{n+1}}$, is the radius of the circle of convergence of the power series.

If we are to define a matrix power series, we must first define the convergence of a sequence of matrices.

Definition 7.17

The sequence of matrices $\{A_r\}$ of order $n \times n$ is said to converge to A, denoted by

$$\lim_{r\to\infty} A_r = A \quad \text{or } \{A_r\} \to A$$

if each of the n^2 sequences of elements converge to the corresponding element of $A = [a_{ij}]$.

For example

$$\lim_{r\to\infty} \begin{bmatrix} \dfrac{2r^2-1}{3r^2+1} & 2 \\ \dfrac{3}{r} & \dfrac{-5r}{r+2} \end{bmatrix} = \begin{bmatrix} \lim_{r\to\infty} \dfrac{2r^2-1}{3r^2+1} & \lim_{r\to\infty} 2 \\ \lim_{r\to\infty} \dfrac{3}{r} & \lim_{r\to\infty} \dfrac{-5r}{r+2} \end{bmatrix} = \begin{bmatrix} \dfrac{2}{3} & 2 \\ 0 & -5 \end{bmatrix}.$$

We could now prove various results for matrix sequences which are analogous to results for sequences of scalar functions. For example, if

$$\{A_r\} \to A \text{ and } \{B_r\} \to B, \text{ then}$$

$\{A_rB_r\} \to AB$ and $\{aA_r+bB_r\} \to aA + bB$ (a, b are scalars).

We leave the interested reader to consult some of the more specialised books mentioned in the Bibliography for rigorous proofs of these and similar results. A particularly important result, which we again state (without proof) is the following:

Theorem 7.17

Let R be the radius of the circle of convergence in which $f(z)$ has the power representation

$$f(z) = \sum_{n=0}^{\infty} a_n z^n, \; |z| < R.$$

If all the eigenvalues λ_i of a matrix A of order $m \times m$ are such that

$$|\lambda_i| < R \quad (i = 1, 2, \ldots m)$$

then the matrix power series

$$\sum_{n=0}^{\infty} a_n A^n \text{ converges to } f(A).$$

(Note that A^0 is defined as I).

For example, since

$$e^z = \sum_{0}^{\infty} \frac{1}{n!} z^n \quad (\text{all } z, \text{ that is, } R = \infty)$$

it follows that (the **matrix exponential**)

$$e^A = \sum_{0}^{\infty} \frac{1}{n!} A^n = I + A + \frac{1}{2!} A^2 + \frac{1}{3!} A^3 + \ldots \qquad (7.35)$$

for every square matrix A.

Example 7.18

Find the matrix power series for the function

$$ln(I + A).$$

Solution

Since

$$ln(1+z) = z - \frac{1}{2}z^2 + \frac{1}{3}z^3 - \ldots \text{ for } |z| < 1$$

it follows that

$$ln(I + A) = A - \frac{1}{2}A^2 + \frac{1}{3}A^3 - \ldots$$

so long as all the eigenvalues of A are such that $|\lambda_i| < 1$ (all i).

The matrix power series of the type (7.35) is useful because we can simplify calculations by making use of an annihilating polynomial (in particular the minimum polynomial) of A.

Example 7.19

Evaluate e^{At} where $A = \begin{bmatrix} 1 & 0 & 0 \\ -1 & 0 & 0 \\ 1 & 1 & 1 \end{bmatrix}$.

Solution

$$e^{At} = I + At + \frac{1}{2!}A^2t^2 + \ldots .$$

In Ex. 6.12 we have found the minimum polynomial for A; it is $\lambda^2 - \lambda = 0$ so that

$$A^2 = A, \text{ hence } A^3 = A, A^4 = A \ \ldots .$$

It follows that

$$e^{At} = I + A(t + \frac{1}{2!}t^2 + \ldots) = I + A(e^t - 1)$$

so that

$$e^{At} = \begin{bmatrix} 1 & 0 & 0 \\ 0 & 1 & 0 \\ 0 & 0 & 1 \end{bmatrix} + \begin{bmatrix} e^t-1 & 0 & 0 \\ -e^t+1 & 0 & 0 \\ e^t-1 & e^t-1 & e^t-1 \end{bmatrix} = \begin{bmatrix} e^t & 0 & 0 \\ 1-e^t & 1 & 0 \\ e^t-1 & e^t-1 & e^t \end{bmatrix}.$$

The fact that so many matrix function relations are analogous to the corresponding scalar function relations must not delude us that this is always so. For example, since

$$e^x e^y = e^{x+y}$$

it does not follow that the corresponding relation with matrix exponentials is always true. Indeed

$$\begin{aligned} e^A e^B &= (I + A + \frac{1}{2!}A^2 + \ldots)(I + B + \frac{1}{2!}B^2 + \ldots) \\ &= I + (A + B) + \frac{1}{2!}(A^2 + 2AB + B^2) + \ldots \end{aligned}$$

and

$$e^{A+B} = I + (A + B) + \frac{1}{2!}(A^2 + AB + BA + B^2) + \ldots .$$

Hence

$$e^A e^B - e^{A+B} = \frac{1}{2!}(AB-BA) + \text{terms involving higher powers of } (AB-BA).$$

So $e^A e^B \neq e^{A+B}$ unless A and B commute.

In the case when the matrix A can be reduced to a diagonal form, the calculation of e^{At} may be greatly simplified.

We have seen (Sec. 7.2) that if A has eigenvalues $\lambda_1, \lambda_2, \ldots \lambda_n$, and $P^{-1}AP = \text{diag}\,\{\lambda_1, \lambda_2, \ldots \lambda_r\} = \Lambda$, then $A^r = P\Lambda^r P^{-1}$ $(r = 1, 2, \ldots)$ Hence if $f(z)$ satisfies the conditions of Theorem 7.17, then

$$f(A) = P\,\text{diag}\,\{f(\lambda_1), f(\lambda_2), \ldots f(\lambda_n)\}\,P^{-1}.$$

(compare this with the result (7.15) of Theorem 7.13).

Hence if $f(A) = e^{At}$ and A can be reduced to a diagonal form, then

$$e^{At} = P\,\text{diag}\,\{e^{\lambda_1 t}, e^{\lambda_2 t}, \ldots e^{\lambda_n t}\}\,P^{-1}. \tag{7.36}$$

Example 7.20

Use the two methods discussed above to evaluate e^{At} when

$$A = \begin{bmatrix} 3 & 4 \\ -2 & -3 \end{bmatrix}.$$

Solution

Method 1
Since $|\lambda I - A| = \lambda^2 - 1$, it follows that

$$A^2 = I,$$

hence

$$\begin{aligned} e^{At} &= I + At + \frac{1}{2!}A^2t^2 + \frac{1}{3!}A^3t^3 + \ldots \\ &= I + At + \frac{1}{2!}t^2 I + \frac{1}{3!}t^3 A + \ldots \\ &= (1 + \frac{1}{2!}t^2 + \ldots)I + (t + \frac{1}{3!}t^3 + \ldots)A \\ &= \frac{1}{2}(e^t + e^{-t})I + \frac{1}{2}(e^t - e^{-t})A \end{aligned}$$

$$= \begin{bmatrix} \frac{1}{2}(e^t+e^{-t}) & 0 \\ 0 & \frac{1}{2}(e^t+e^{-t}) \end{bmatrix} + \begin{bmatrix} \frac{3}{2}(e^t-e^{-t}) & 2(e^t-e^{-t}) \\ -(e^t-e^{-t}) & -\frac{3}{2}(e^t-e^{-t}) \end{bmatrix}$$

$$= \begin{bmatrix} 2e^t-e^{-t} & 2(e^t-e^{-t}) \\ e^{-t}-e^t & 2e^{-t}-e^t \end{bmatrix}.$$

Method 2
The eigenvalues of A are $\lambda_1 = 1$ and $\lambda_2 = -1$. The modal matrix is

$$P = \begin{bmatrix} 2 & 1 \\ -1 & -1 \end{bmatrix} \text{ and } P^{-1} = \begin{bmatrix} 1 & 1 \\ -1 & -2 \end{bmatrix}$$

so that by (7.36)

$$e^{At} = \begin{bmatrix} 2 & 1 \\ -1 & -1 \end{bmatrix} \begin{bmatrix} e^t & 0 \\ 0 & e^{-t} \end{bmatrix} \begin{bmatrix} 1 & 1 \\ -1 & -2 \end{bmatrix}$$

$$= \begin{bmatrix} 2e^t-e^{-t} & 2(e^t-e^{-t}) \\ e^{-t}-e^t & 2e^{-t}-e^t \end{bmatrix}.$$

In Sec. 1.5 the derivative of a vector was defined – at least implicitly – in terms of its elements. We now define explicitly the derivative and integral of a matrix $A(t)$.

Definition 7.18
Let $A(t) = [a_{ij}(t)]$, then

(1) $\frac{d}{dt}A(t) = \dot{A}(t) = [\frac{d}{dt}(a_{ij})]$ and

(2) $\int A(t)dt = [\int a_{ij}\,dt]$.

For example if

$$A = \begin{bmatrix} 6t & \sin 2t \\ t^2+2 & 3 \end{bmatrix}, \text{ then}$$

$$\dot{A} = \begin{bmatrix} 6 & 2\cos 2t \\ 2t & 0 \end{bmatrix}, \text{ and}$$

$$\int A\,dt = \begin{bmatrix} 3t^2 & -\frac{1}{2}\cos 2t \\ \frac{1}{3}t^3+2t & 3t \end{bmatrix} + C$$

where C is a constant matrix (of integration).

From the definition, it follows that

$$\frac{d}{dt}(aA + bB) = a\dot{A} + b\dot{B}$$

$$\int_\alpha^\beta (aA + bB)dt = a\int_\alpha^\beta A\,dt + b\int_\alpha^\beta B\,dt$$

for all constants a, b.

Also it can be shown that

$$\frac{d}{dt}(AB) = \dot{A}B + A\dot{B}, \text{ and if } A \text{ is a constant matrix,}$$

$$\frac{d}{dt}(c^{At}) = Ae^{At}.$$

But it is *not* true (in general) that

$$\frac{d}{dt}(A^n) = nA^{n-1}\dot{A}.$$

Indeed

$$\frac{d}{dt}(A^n) = \frac{d}{dt}(A.A^{n-1}) = \left[\frac{d}{dt}(A)\right]A^{n-1} + A\frac{d}{dt}(A^{n-1})$$

$$= \left[\frac{d}{dt}(A)\right].A^{n-1} + A\,\frac{d}{dt}(A)\,A^{n-2} + A^2\frac{d}{dt}(A^{n-2})$$

$$= \left[\frac{d}{dt}(A)\right].A^{n-1} + A\left[\frac{d}{dt}(A)\right]A^{n-2} + \ldots + A^{n-1}\frac{d}{dt}(A).$$

But since, in general, A and $\dot{A}$ do not commute, this cannot be simplified.

We are now in a position to discuss the solution of state equations, first introduced in Sec. 1.5,

$$\begin{aligned} \dot{\mathbf{x}} &= A\mathbf{x} + Bu \\ y &= C\mathbf{x} \end{aligned} \tag{7.37}$$

We first consider the homogeneous case, that is

$$\dot{\mathbf{x}} = A\mathbf{x} \tag{7.38}$$

with initial conditions $\mathbf{x}(0) = \mathbf{x}_0$.

As in the scalar case, we verify that

$$\mathbf{x} = e^{At}\mathbf{x}(0) \tag{7.39}$$

is a solution.

Indeed

$$\dot{\mathbf{x}} = Ae^{At}\mathbf{x}(0) = A\mathbf{x}$$

and

$$\mathbf{x}(0) = e^{0}\mathbf{x}(0) = I\mathbf{x}(0) = \mathbf{x}(0).$$

The matrix e^{At} in the solution (7.39) is of special interest to control engineers; they call it the **state-transition matrix** (it is a transition matrix as defined in Sec. 2.3) and denote it by $\Phi(t)$, that is

$$\Phi(t) = e^{At} \quad .$$

Notice that $\Phi(t)$ is a linear transformation of the state of the system $\mathbf{x}(0)$ at some initial time $t = 0$, to the state of the system $\mathbf{x}(t)$ at some subsequent time t, hence the name given to this matrix.

Since

$$e^{A}e^{-A} = I,$$

it follows that

$$(e^{A})^{-1} = e^{-A}.$$

Hence $\Phi^{-1}(t) = e^{-At} = \Phi(-t)$.
Also $\Phi(t)\,\Phi(-y) = e^{At}\,e^{-Ay} = e^{A(t-y)} = \Phi(t-y)$. (7.40)

Writing (7.37) in the form

$$\dot{\mathbf{x}} - A\mathbf{x} = Bu,$$

and multiplying by e^{-At}, we obtain

$$e^{-At}\,(\dot{\mathbf{x}} - A\mathbf{x}) = e^{-At}Bu$$

or,

$$\frac{d}{dt}\,[e^{-At}\,\mathbf{x}] = e^{-At}Bu.$$

Integrating between 0 and t, the above becomes

$$e^{-At}\,\mathbf{x}(t) - \mathbf{x}(0) = \int_0^t e^{-A\tau}\,Bu(\tau)d\tau$$

so that

$$\mathbf{x}(t) = e^{At}\mathbf{x}(0) + e^{At}\int_0^t e^{-A\tau}Bu(\tau)d\tau \quad \text{and by (7.40)}$$

$$= \Phi(t)\,\mathbf{x}(0) + \int_0^t \Phi(t-\tau)\,B\tilde{u}\,(\tau)d\tau \ . \tag{7.41}$$

Example 7.21
Obtain the state $\mathbf{x}(t)$ at time t of the system specified by (see Sec. 1.5)

$$\begin{bmatrix} \dot{x}_1 \\ \dot{x}_2 \end{bmatrix} = \begin{bmatrix} 0 & 2 \\ -1 & -3 \end{bmatrix} \begin{bmatrix} x_1 \\ x_2 \end{bmatrix} + \begin{bmatrix} 0 \\ 1 \end{bmatrix} [u]$$

given (1) $u(t) = 1$ for $t \geq 0$, and (2) the initial state of the system $\mathbf{x}(0) = \begin{bmatrix} 1 \\ -1 \end{bmatrix}$.

Solution

The matrix $A = \begin{bmatrix} 0 & 2 \\ -1 & -3 \end{bmatrix}$ has eigenvalues $\lambda_1 = -1$ and $\lambda_2 = -2$.

The corresponding modal matrix is

$$P = \begin{bmatrix} 2 & 1 \\ -1 & -1 \end{bmatrix} \quad \text{so that } P^{-1} = \begin{bmatrix} 1 & 1 \\ -1 & -2 \end{bmatrix}$$

hence

$$\Phi(t) = e^{At} = \begin{bmatrix} 2 & 1 \\ -1 & -1 \end{bmatrix} \begin{bmatrix} e^{-t} & 0 \\ 0 & e^{-2t} \end{bmatrix} \begin{bmatrix} 1 & 1 \\ -1 & -2 \end{bmatrix}$$

$$= \begin{bmatrix} 2e^{-t}-e^{-2t} & 2e^{-t}-2e^{-2t} \\ e^{-2t}-e^{-t} & 2e^{-2t}-e^{-t} \end{bmatrix}$$

and

$$\Phi(t)\mathbf{x}(0) = e^{-2t}\begin{bmatrix} 1 \\ -1 \end{bmatrix}$$

Since $B = \begin{bmatrix} 0 \\ 1 \end{bmatrix}$

$$\Phi(t-\tau)Bu = \begin{bmatrix} 2e^{-(t-\tau)}-2e^{-2(t-\tau)} \\ 2e^{-2(t-\tau)}-e^{-(t-\tau)} \end{bmatrix}$$

and

$$\int_0^t \phi(t-\tau)Bu \, d\tau = \begin{bmatrix} 1-2e^{-t}+e^{-2t} \\ e^{-t}-e^{-2t} \end{bmatrix}.$$

Hence, at time t the response of the system is

$$\mathbf{x}(t) = \Phi(t)\mathbf{x}(0) + \int_0^t \Phi(t-\tau)Bu \, d\tau = \begin{bmatrix} 1-2e^{-t}+2e^{-2t} \\ e^{-t}-2e^{-2t} \end{bmatrix}.$$

PROBLEMS FOR CHAPTER 7

1) Given $f(x) = 2x^2 - x + 1$, find $f(A)$ when

(i)
$$A = \begin{bmatrix} 1 & 1 & 0 \\ 0 & 1 & 0 \\ 0 & 0 & 1 \end{bmatrix};$$

(ii)
$$A = \begin{bmatrix} 0 & 1 & 0 \\ 1 & 0 & 0 \\ 0 & 0 & 1 \end{bmatrix}.$$

In both cases calculate $|f(A)|$ by two different methods.

2) Given the matrix A in Problem 1(ii) above, calculate by two different methods the eigenvalues of $f(A)$, when

(i) $f(x) = 3x^2 - 2x + 2;$

(ii) $f(x) = x^2 + 2x + 5.$

3) Find the characteristic equation and the minimal polynomial for

(i)
$$A = \begin{bmatrix} 1 & 1 & -2 \\ -1 & 2 & 1 \\ 0 & 1 & -1 \end{bmatrix};$$

(ii)
$$A = \begin{bmatrix} 0 & 0 & 0 \\ 1 & 0 & 0 \\ 0 & 0 & 0 \end{bmatrix}.$$

4) Consider the matrix

$$A = \begin{bmatrix} 3 & 0 & 0 \\ 1 & 3 & 0 \\ 2 & -1 & -2 \end{bmatrix}.$$

(i) Find the characteristic equation and the minimal polynomial for A.
(ii) State whether A can be reduced by a similarity transformation to a diagonal form – give reasons.
(iii) Find the A-invariant subspaces associated with A.
(iv) Select cyclic bases for the subspaces in (iii).
(v) Find a matrix P such that $P^{-1}AP$ is in a block diagonal form.
(vi) Write A in the form $A_1 \oplus A_2$.

5) Consider the matrix

$$A = \begin{bmatrix} 8 & 10 & -3 \\ 0 & 2 & 0 \\ 18 & 29 & -7 \end{bmatrix}.$$

(i) Find the minimal polynomial for A in factored form.
(ii) Write down the (lower) Jordan canonical form corresponding to each factor.
(iii) Write down the (lower) Jordan block form of the matrix A.
(iv) Verify the result by finding the matrix P such that $P^{-1}AP$ is in the Jordan block form.

6) Consider the matrix

$$A = \begin{bmatrix} 2 & 1 & -2 \\ 0 & -1 & 1 \\ 0 & 0 & -1 \end{bmatrix}.$$

Find the matrix P such that $P^{-1}AP$ is in (upper) Jordan block form.

7) Evaluate e^{At} when

(i) $$A = \begin{bmatrix} 0 & 1 & 0 \\ 1 & 0 & 0 \\ 0 & 0 & 1 \end{bmatrix};$$

(ii) $$A = \begin{bmatrix} 1 & 1 & 1 & 1 \\ 0 & -1 & -2 & -3 \\ 0 & 0 & 1 & 3 \\ 0 & 0 & 0 & -1 \end{bmatrix};$$

(iii)

$$A = \begin{bmatrix} 1 & 5 & -2 \\ 1 & 2 & -1 \\ 3 & 6 & -3 \end{bmatrix}.$$

8) Find the general solution of

$$\begin{bmatrix} \dot{x}_1 \\ \dot{x}_2 \end{bmatrix} = \begin{bmatrix} 6 & 6 \\ -12 & -11 \end{bmatrix} \begin{bmatrix} x_1 \\ x_2 \end{bmatrix} + \begin{bmatrix} 0 \\ 1 \end{bmatrix} [u] ,$$

when $u(t) = 1$ for $t \geq 0$ and $\mathbf{x}(0) = \begin{bmatrix} 0 \\ 1 \end{bmatrix}$.

CHAPTER 8

Inverting a Matrix

In this chapter we discuss a number of 'practical' methods for inverting certain types of matrices. The 'practical' does not necessarily refer to numerical methods (see Ref [10]) but to methods which are analytical and useful for both numerical evaluations and as 'closed formulae' in theoretical discussions.

An indication of the importance of various methods for inverting matrices in Control Engineering can perhaps be appreciated when we consider again the system equations

$$\dot{\mathbf{x}} = A\mathbf{x} + B\mathbf{u}$$

$$\mathbf{y} = C\mathbf{x}.$$

On taking the Laplace transforms, we obtain

$$s\mathbf{X} = A\mathbf{X} + B\mathbf{U}$$

$$\mathbf{Y} = C\mathbf{X}$$

(where **X**, **U** and **Y** are the Laplace transforms of **x**, **u** and **y** respectively).

Hence

$$(sI - A)\mathbf{X} = B\mathbf{U}$$

or

$$\mathbf{X} = (sI - A)^{-1}B\mathbf{U}$$

and so

$$\mathbf{Y} = C(sI - A)^{-1}B\mathbf{U}.$$

The matrix $C(sI - A)^{-1}B$, called the **transfer function matrix**, relates the input and the output of the system and plays a crucial role in control theory.

This function has a factor which is the inverse of the matrix $(sI - A)$. For an understanding of the dynamic behaviour of the system, under different conditions, various methods for inverting matrices, some of them discussed in this chapter, are used.

The first technique we discuss can be used to both invert a matrix and also to determine the rank of a matrix.

8.1 ELEMENTARY OPERATIONS AND ELEMENTARY MATRICES

Basically there are three operations on matrices known as **Elementary** or E-operations.

Definition 8.1

(1) The interchange of two rows (or columns).
(2) The multiplication of a row (or column) by a scalar k.
(3) The addition of a multiple k of one row (or column) to another row (or column).

The notation used is the following: Consider a matrix having 3 rows, R_1, R_2 and R_3 and 3 columns C_1, C_2 and C_3. The matrix need not be square; k is a constant.

(1) Interchange R_1 and R_2.

$$R_1 \leftrightarrow R_2 \qquad\qquad (\text{or } C_1 \leftrightarrow C_2).$$

(2) Multiply R_1 by k.

$$R_1 \to kR_1 \qquad\qquad (\text{or } C_1 \to kC_1).$$

(3) Add a multiple k of R_1 to R_3.

$$R_3 \to R_3 + kR_1 \qquad\qquad (\text{or } C_3 \to C_3 + kC_1)$$

means the addition of k times R_1 (C_1) to R_3 (C_3).
For example, given

$$A = \begin{bmatrix} a_1 & b_1 & c_1 \\ a_2 & b_2 & c_2 \\ a_3 & b_3 & c_3 \end{bmatrix}$$

then the operation $R_3 \to R_3 + 2R_1$ on A, results in the matrix

$$B = \begin{bmatrix} a_1 & b_1 & c_1 \\ a_2 & b_2 & c_2 \\ a_3+2a_1 & b_3+2b_1 & c_3+2c_1 \end{bmatrix}.$$

One important property of the elementary operations is that each one has an **inverse** which is itself an elementary operation. For example, to obtain A from the matrix B above we need to apply (to B) the elementary operation

$$R_3 \to R_3 - 2R_1.$$

Hence the inverse of

$$R_3 \to R_3 + kR_1 \text{ is } R_3 \to R_3 - kR_1.$$

Another important property is that if the matrix B is the result of applying an elementary operation to A, then

$$r(A) = r(B),$$

that is, the rank of a matrix remains unchanged under elementary operations. This result follows from the discussion in Sec. 4.2. There is no loss of generality if we confine ourselves to elementary row operations only – similar results apply to elementary column operations.

Since the rank of a matrix A equals the number of linearly independent (row) vectors making up the matrix, interchanging two rows, or multiplying a row by a scalar, does not change the rank.

We need therefore only consider whether our elementary operation of type 3 can change the rank of A.

We can argue (non-rigorously) that if $W_1 = \{\mathbf{z}_1, \mathbf{z}_2, \mathbf{z}_3\}$ is the space spanned by the 3 row vectors of A, and $W_2 = \{\mathbf{z}_1, \mathbf{z}_2, \mathbf{z}_3 + k\mathbf{z}_1\}$ is the space spanned by the 3 row vectors of B – the matrix obtained when an elementary operation of type 3 is applied to A, then

$$W_1 = W_2$$

so that $\dim W_1 = \dim W_2$, and

$$r(A) = r(B).$$

Definition 8.2
The matrices A and B are said to be **equivalent**, denoted by $A \sim B$, if one can be obtained from the other by a sequence of elementary operations.

Example 8.1
Show that the matrices

$$A = \begin{bmatrix} 1 & 2 & 0 \\ 3 & 0 & 6 \\ -2 & 1 & -5 \end{bmatrix} \text{ and } B = \begin{bmatrix} 1 & 0 & 0 \\ 0 & 1 & 0 \\ 0 & 0 & 0 \end{bmatrix}$$

are equivalent. Find the rank of A.

Solution
Starting with A we carry out the following elementary operations:

$R_2 \to R_2 - 3R_1, A$ becomes

$$A_1 = \begin{bmatrix} 1 & 2 & 0 \\ 0 & -6 & 6 \\ -2 & 1 & -5 \end{bmatrix}$$

$R_3 \to R_3 + 2R_1, A_1$ becomes

$$A_2 = \begin{bmatrix} 1 & 2 & 0 \\ 0 & -6 & 6 \\ 0 & 5 & -5 \end{bmatrix}$$

$R_2 \to -\dfrac{1}{6}R_2, A_2$ becomes

$$A_3 = \begin{bmatrix} 1 & 2 & 0 \\ 0 & 1 & -1 \\ 0 & 5 & -5 \end{bmatrix}$$

$R_3 \to R_3 - 5R_2, A_3$ becomes

$$A_4 = \begin{bmatrix} 1 & 2 & 0 \\ 0 & 1 & -1 \\ 0 & 0 & 0 \end{bmatrix}$$

$C_2 \to C_2 - 2C_1, A_4$ becomes

$$A_5 = \begin{bmatrix} 1 & 0 & 0 \\ 0 & 1 & -1 \\ 0 & 0 & 0 \end{bmatrix}$$

$C_3 \to C_3 + C_2, A_5$ becomes

$$A_6 = \begin{bmatrix} 1 & 0 & 0 \\ 0 & 1 & 0 \\ 0 & 0 & 0 \end{bmatrix} = B.$$

Hence $A \sim B$.

Since $r(B) = 2, r(A) = 2$.

It is a remarkable fact that an elementary operation can be performed on a matrix A, by multiplying A by an appropriate elementary matrix.

Definition 8.3

A matrix obtained from the unit matrix by an E-operation is called an **elementary** or an E-**matrix.**

For example, corresponding to the E-operations

$$R_1 \leftrightarrow R_3, C_1 \leftrightarrow C_3, \text{and } R_3 \to R_3 - 5R_1$$

on a matrix A of order 3X3, the E-matrices are

$$\begin{bmatrix} 0 & 0 & 1 \\ 0 & 1 & 0 \\ 1 & 0 & 0 \end{bmatrix}, \begin{bmatrix} 0 & 0 & 1 \\ 0 & 1 & 0 \\ 1 & 0 & 0 \end{bmatrix} \text{ and } \begin{bmatrix} 1 & 0 & 0 \\ 0 & 1 & 0 \\ -5 & 0 & 1 \end{bmatrix}.$$

Notice that since $R_i \leftrightarrow R_j$ and $C_i \leftrightarrow C_j$ have the same effect on I, we expect the E-matrices corresponding to $R_1 \leftrightarrow R_3$ and $C_1 \leftrightarrow C_3$ to be the same. Similarly the E-matrices corresponding to the E-operation $R_i \rightarrow kR_i$ and $C_i \rightarrow kC_i$ are the same.

On the other hand the E-matrix corresponding to $R_i \rightarrow R_i + kR_j$ is *different* from the E-matrix corresponding to $C_i \rightarrow C_i + kC_j$. But the E-matrices corresponding to $R_i \rightarrow R_i + kR_j$ and $C_j \rightarrow C_j + kC_i$ are the same. For example the E-matrix of order 3X3 corresponding to the operations

$R_3 \rightarrow R_3 + kR_1$ and $C_1 \rightarrow C_1 + kC_3$ is

$$\begin{bmatrix} 1 & 0 & 0 \\ 0 & 1 & 0 \\ k & 0 & 1 \end{bmatrix}$$

From considerations of matrix multiplication we can easily deduce that:

(1) Premultiplication of a matrix A by an E-matrix is equivalent to the corresponding E-operation on the rows of A.

(2) Postmultiplication of A by an E-matrix, is equivalent to the corresponding E-operation on the columns of A.

Example 8.2

Given the matrix

$$A = \begin{bmatrix} a_1 & a_2 & a_3 \\ b_1 & b_2 & b_3 \\ c_1 & c_2 & c_3 \end{bmatrix},$$

use appropriate E-matrices to carry out the operations

(1) $R_3 \rightarrow R_3 - 5R_1$ and (2) $C_3 \rightarrow C_3 - 5C_1$ on A.

Solution

(1) We premultiply A by the appropriate E-matrix; that is,

$$\begin{bmatrix} 1 & 0 & 0 \\ 0 & 1 & 0 \\ -5 & 0 & 1 \end{bmatrix} \begin{bmatrix} a_1 & a_2 & a_3 \\ b_1 & b_2 & b_3 \\ c_1 & c_2 & c_3 \end{bmatrix} = \begin{bmatrix} a_1 & a_2 & a_3 \\ b_1 & b_2 & b_3 \\ c_1-5a_1 & c_2-5a_2 & c_3-5a_3 \end{bmatrix}$$

which corresponds to $R_3 \rightarrow R_3 - 5R_1$.

(2) We postmultiply A by the appropriate E-matrix

$$\begin{bmatrix} a_1 & a_2 & a_3 \\ b_1 & b_2 & b_3 \\ c_1 & c_2 & c_3 \end{bmatrix} \begin{bmatrix} 1 & 0 & -5 \\ 0 & 1 & 0 \\ 0 & 0 & 1 \end{bmatrix} = \begin{bmatrix} a_1 & a_2 & a_3-5a_1 \\ b_1 & b_2 & b_3-5b_1 \\ c_1 & c_2 & c_3-5c_1 \end{bmatrix}$$

which corresponds to $C_3 \rightarrow C_3 - 5C_1$.

Corresponding to the properties of E-operations discussed above, an E-matrix has the following ones:

(1) It is non-singular

(2) Its inverse is an E-matrix.

Property (1) is obvious, since an E-matrix is the result of an E-operation on a unit matrix. Since an E-operation leaves the rank of the unit matrix unchanged, the corresponding E-matrix is non-singular (that is, its rank is the same as that of the unit matrix).

To verify (2) we note that since each elementary operation has an inverse which is also an elementary operation, the inverse of an E-matrix is another E-matrix.

For example, since the operation $R_3 \rightarrow R_3 + kR_1$ has the inverse $R_3 \rightarrow R_3 - kR_1$, the two corresponding E-matrices (of order 3X3)

$$E_1 = \begin{bmatrix} 1 & 0 & 0 \\ 0 & 1 & 0 \\ k & 0 & 1 \end{bmatrix} \text{ and } E_2 = \begin{bmatrix} 1 & 0 & 0 \\ 0 & 1 & 0 \\ -k & 0 & 1 \end{bmatrix}$$

are inverses of each other, as can be verified by noting that $E_1E_2 = I$.

We can now make use of E-matrices to find the inverse of a non-singular matrix A.

Since A is equivalent to the unit matrix (of appropriate order) we need to find a sequence of E-matrices, say $E_1, E_2, \ldots E_s$ such that

$$(E_sE_{s-1} \ldots E_2E_1)A = I$$

so that

$$A^{-1} = E_sE_{s-1} \ldots E_2E_1.$$

(Also see Problem 3 at the end of this chapter.)

Example 8.3

Find the inverse of the matrix

$$A = \begin{bmatrix} 1 & 0 & -2 \\ 4 & 1 & 0 \\ 0 & 2 & 1 \end{bmatrix}.$$

Solution

To reduce A to a unit matrix, we can use the following E-operations

$R_2 \to R_2 - 4R_1$ corresponding to E-matrix

$$E_1 = \begin{bmatrix} 1 & 0 & 0 \\ -4 & 1 & 0 \\ 0 & 0 & 1 \end{bmatrix}$$

$R_3 \to R_3 - 2R_2$ corresponding to

$$E_2 = \begin{bmatrix} 1 & 0 & 0 \\ 0 & 1 & 0 \\ 0 & -2 & 1 \end{bmatrix}$$

$R_3 \to -\frac{1}{15} R_3$ corresponding to

$$E_3 = \begin{bmatrix} 1 & 0 & 0 \\ 0 & 1 & 0 \\ 0 & 0 & -\frac{1}{15} \end{bmatrix}$$

$R_1 \to R_1 + 2R_3$ corresponding to

$$E_4 = \begin{bmatrix} 1 & 0 & 2 \\ 0 & 1 & 0 \\ 0 & 0 & 1 \end{bmatrix}$$

and finally,

$R_2 \to R_2 - 8R_3$ corresponding to

$$E_5 = \begin{bmatrix} 1 & 0 & 0 \\ 0 & 1 & -8 \\ 0 & 0 & 1 \end{bmatrix}.$$

Since these 5 E-operations reduce A to the unit matrix, it follows that

$$A^{-1} = E_5E_4E_3E_2E_1$$

$$= \frac{1}{15}\begin{bmatrix} -1 & 4 & -2 \\ 4 & -1 & 8 \\ -8 & 2 & -1 \end{bmatrix}.$$

For a more direct method see Problem 3 at the end of this chapter.

8.2 THE INVERSE OF A VANDERMONDE MATRIX

We have made use of companion matrices in the discussion of reducing a matrix to a canonical form in Chapter 7.

In fact this particular form of a matrix is found useful in various applications in engineering, some of them being in control theory, in particular when considering certain feedback laws.

In what follows we shall assume that the eigenvalues of the matrices to be considered are distinct. At first sight this assumption may seem a little unreasonable, since we have spent so much space in this book in discussing the case when the eigenvalues are not necessarily distinct. In fact it is very important that we understand the theoretical consequences of non-distinct eigenvalues, even if for practical systems we can claim that the eigenvalues are distinct, although some of them can be 'very near' to each other.

Given a matrix A, having distinct eigenvalues $\lambda_1, \lambda_2, \ldots \lambda_n$, the corresponding Vandermonde matrix has the form:

$$V = \begin{bmatrix} 1 & 1 & \dots & 1 \\ \lambda_1 & \lambda_2 & & \lambda_n \\ \lambda_1^2 & \lambda_2^2 & & \lambda_n^2 \\ \vdots & \vdots & & \vdots \\ \lambda_1^{n-1} & \lambda_2^{n-1} & & \lambda_n^{n-1} \end{bmatrix}. \tag{8.1}$$

The matrix V is closely associated with the matrix A in the companion form

$$A = \begin{bmatrix} 0 & 1 & 0 & \dots & 0 \\ 0 & 0 & 1 & \dots & 0 \\ \vdots & & & & \\ 0 & 0 & 0 & \dots & 1 \\ -a_0 & -a_1 & -a_2 & \dots & -a_{n-1} \end{bmatrix} \tag{8.2}$$

where the characteristic equation of A is

$$|\lambda I - A| = \lambda^n + a_{n-1}\lambda^{n-1} + \dots + a_1\lambda + a_0 = 0.$$

(Notice that (8.2) is the transposed form of the matrix (7.28); it is still called a companion matrix).

The association between (8.1) and (8.2) is easily verified to be

$$V^{-1}AV = \Lambda = \text{diag}\{\lambda_1, \lambda_2, \dots \lambda_n\} \tag{8.3}$$

so that V is the modal matrix for A.

To apply the similarity transformation (8.3) we need to calculate V^{-1}.

The method we shall discuss makes use of the **Lagrange interpolation polynomial** (see [10])

$$f(z) = \sum_{i=1}^{n} f(\lambda_i)L_i(z) \tag{8.4}$$

where

$$L_i(z) = \frac{(z-\lambda_1)\ \ldots\ (z-\lambda_{i-1})(z-\lambda_{i+1})\ \ldots\ (z-\lambda_n)}{(\lambda_i-\lambda_1)\ \ldots\ (\lambda_i-\lambda_{i-1})(\lambda_i-\lambda_{i+1})\ \ldots\ (\lambda_i-\lambda_n)} \tag{8.5}$$

for $i = 1, 2, \ldots n$.

The function f (Eq. 8.4) is the polynomial of degree $(n-1)$ whose images at $z = \lambda_1, \lambda_2, \ldots \lambda_n$ are $f(\lambda_1), f(\lambda_2), \ldots f(\lambda_n)$.

In particular, if we choose

$$f(z) = z^k \quad (k = 0, 1, \ldots n-1)$$

then for

$$k = 0,\ \ 1 = L_1 + L_2 + \ldots + L_n$$

$$k = 1,\ \ z = \lambda_1 L_1 + \lambda_2 L_2 + \ldots + \lambda_n L_n$$

$$\vdots$$

$$k = n-1,\ \ z^{n-1} = \lambda_1^{n-1} L_1 + \lambda_2^{n-1} L_2 + \ldots + \lambda_n^{n-1} L_n,$$

(where L_i stands for $L_i(z)$.) We write the above set of equations in matrix form, as

$$\mathbf{Z} = V\mathbf{L} \tag{8.6}$$

where

$$\mathbf{Z} = \begin{bmatrix} 1 \\ z \\ \vdots \\ z^{n-1} \end{bmatrix} \quad \mathbf{L} = \begin{bmatrix} L_1 \\ L_2 \\ \vdots \\ L_n \end{bmatrix}, \text{ and } V = \begin{bmatrix} 1 & 1 & \ldots & 1 \\ \lambda_1 & \lambda_2 & \ldots & \lambda_n \\ \vdots & & & \\ \lambda_1^{n-1} & \lambda_2^{n-1} & \ldots & \lambda_n^{n-1} \end{bmatrix}$$

From (8.6) we have

$$\mathbf{L} = V^{-1}\mathbf{Z}. \tag{8.7}$$

Assume that

$$V^{-1} = \begin{bmatrix} \alpha_{11} & \alpha_{12} & \dots & \alpha_{1n} \\ \alpha_{21} & \alpha_{22} & \dots & \alpha_{2n} \\ \vdots & & & \\ \alpha_{n1} & \alpha_{n2} & \dots & \alpha_{nn} \end{bmatrix}$$

then (8.7) becomes

$$\begin{bmatrix} L_1 \\ L_2 \\ \vdots \\ L_n \end{bmatrix} = \begin{bmatrix} \alpha_{11} & \alpha_{12} & \dots & \alpha_{1n} \\ \alpha_{21} & \alpha_{22} & \dots & \alpha_{2n} \\ \vdots & & & \\ \alpha_{n1} & \alpha_{n2} & & \alpha_{nn} \end{bmatrix} \begin{bmatrix} 1 \\ z \\ \vdots \\ z^{n-1} \end{bmatrix}$$

that is,

$$L_i = \alpha_{i1} + \alpha_{i2} z + \dots + \alpha_{in} z^{n-1} \qquad (i = 1, 2, \dots n). \tag{8.8}$$

Eq. (8.8) shows that the $(i, j)^{\text{th}}$ element α_{ij} of the matrix V^{-1} is the coefficient of z^{j-1} in the polynomial L_i.

Example 8.4

Consider $A = \begin{bmatrix} 0 & 1 & 0 \\ 0 & 0 & 1 \\ 2 & -1 & 2 \end{bmatrix}$, whose characteristic equation is

$$C(\lambda) = \lambda^3 - 2\lambda^2 + \lambda - 2 = 0$$

and characteristic roots are

$$\lambda_1 = 2, \ \lambda_2 = i \text{ and } \lambda_3 = -i.$$

The corresponding Vandermonde matrix is

$$V = \begin{bmatrix} 1 & 1 & 1 \\ 2 & i & -i \\ 4 & -1 & -1 \end{bmatrix}.$$

Find V^{-1}.

Solution
By (8.5) we obtain

$$L_1(z) = \frac{(z-\lambda_2)(z-\lambda_3)}{(\lambda_1-\lambda_2)(\lambda_1-\lambda_3)} = \frac{1 + z^2}{5}$$

so that $\alpha_{11} = \frac{1}{5}$, $\alpha_{12} = 0$, and $\alpha_{13} = \frac{1}{5}$.

$$L_2(z) = \frac{(z-\lambda_1)(z-\lambda_3)}{(\lambda_2-\lambda_1)(\lambda_2-\lambda_3)} = \frac{(4+2i) - 5iz + (2i-1)z^2}{10}$$

so that $\alpha_{21} = \frac{4+2i}{10}$, $\alpha_{22} = \frac{-5i}{10}$ and $\alpha_{33} = \frac{2i-1}{10}$.

$$L_3(z) = \frac{(z-\lambda_1)(z-\lambda_2)}{(\lambda_3-\lambda_1)(\lambda_3-\lambda_2)} = \frac{(8-4i) + 10iz - (2+4i)z^2}{20}$$

so that $\alpha_{31} = \frac{8-4i}{20}$, $\alpha_{32} = \frac{10i}{20}$ and $\alpha_{33} = \frac{-(2+4i)}{20}$.

Hence,

$$V^{-1} = \frac{1}{20}\begin{bmatrix} 4 & 0 & 4 \\ 8+4i & -10i & -2+4i \\ 8-4i & 10i & -2-4i \end{bmatrix}.$$

8.3 FADDEEVA'S METHOD

This method (also known as the Leverrier's Method), although rather slow, has the advantage of producing simultaneously both the inverse and the characteristic equation of a non-singular matrix A (of order $n \times n$).

The method makes use of an algorithm which defines a sequence of matrices, the penultimate member of the sequence being a scalar multiple of A^{-1}.

Before defining the algorithm we discuss a preliminary result.

The matrix A is assumed to have the characteristic equation

$$C(\lambda) = \lambda^n + b_1\lambda^{n-1} + \ldots + b_{n-1}\lambda + b_n = 0 \tag{8.9}$$

and characteristic roots $\lambda_1, \lambda_2, \ldots \lambda_n$, so that

$$C(\lambda) = (\lambda-\lambda_1)(\lambda-\lambda_2)\ldots(\lambda-\lambda_n). \tag{8.10}$$

Hence

$$C'(\lambda) = n\lambda^{n-1} + (n-1)b_1\lambda^{n-2} + \ldots + b_{n-1}, \tag{8.11}$$

also

$$C'(\lambda) = \frac{C(\lambda)}{\lambda-\lambda_1} + \frac{C(\lambda)}{\lambda-\lambda_2} + \ldots + \frac{C(\lambda)}{\lambda-\lambda_n}. \tag{8.12}$$

But (on division)

$$\frac{C(\lambda)}{\lambda-\lambda_1} = \lambda^{n-1} + (\lambda_1+b_1)\lambda^{n-2} + (\lambda_1^2+b_1\lambda_1+b_2)\lambda^{n-2} + \ldots + (\lambda_1^{n-1}+b_1\lambda_1^{n-2} + \ldots + b_n)$$

$$\frac{C(\lambda)}{\lambda-\lambda_2} = \lambda^{n-1} + (\lambda_2+b_1)\lambda^{n-2} + (\lambda_2^2+b_1\lambda_2+b_2)\lambda^{n-2} + \ldots + (\lambda_2^{n-1}+b_1\lambda_2^{n-2} + \ldots + b_n)$$

$$\vdots$$

$$\frac{C(\lambda)}{\lambda-\lambda_n} = \lambda^{n-1} + (\lambda_n+b_1)\lambda^{n-2} + (\lambda_n^2+b_1\lambda_n+b_2)\lambda^{n-2} + \ldots + (\lambda_n^{n-1}+b_1\lambda_n^{n-2} + \ldots + b_n).$$

On addition, and making use of (8.12), we obtain

$$C'(\lambda) = n\lambda^{n-1} + (S_1+nb_1)\lambda^{n-2} + (S_2+b_1S_1+nb_2)\lambda^{n-2} + \ldots + (S_{n-1}+b_1S_{n-2} + \ldots + nb_n) \qquad (8.13)$$

where $S_i = \sum_{r=1}^{n} \lambda_r^i \quad (i = 1, 2, \ldots n-1)$.

Equating the coefficients in (8.11) and (8.13), we obtain

$$(n-1)b_1 = S_1 + nb_1$$

$$(n-2)b_2 = S_2 + b_1S_1 + nb_2$$

and so on. It follows that

$$-b_1 = S_1$$

$$-2b_2 = S_2 + b_1S_1$$

and in general, for $r = 1, 2, \ldots n-1$

$$-rb_r = S_r + b_1S_{r-1} + \ldots + b_{r-1}S_1. \qquad (8.14)$$

We know (see Eqs (6.5) and (6.6)) that

$$C(0) = b_n = (-1)^n \lambda_1\lambda_2 \ldots \lambda_n \qquad (8.15)$$

$$-b_1 = \text{tr}\, A = \lambda_1 + \lambda_2 + \ldots + \lambda_n \qquad (8.16)$$

and (see Sec. 7.2)

$$\text{tr}\, A^i = \lambda_1^i + \lambda_2^i + \ldots + \lambda_n^i \quad (i = 1, 2, \ldots n). \qquad (8.17)$$

We can now define the algorithm.

Let

$$
\begin{aligned}
A_1 &= A, \operatorname{tr} A_1 = -t_1 \text{ and } B_1 = A_1 + t_1 I \\
A_2 &= AB_1, \frac{1}{2}\operatorname{tr} A_2 = -t_2, B_2 = A_2 + t_2 I \\
&\vdots \\
A_{n-1} &= AB_{n-2}, \frac{1}{n-1}\operatorname{tr} A_{n-1} = -t_{n-1}, B_{n-1} = A_{n-1} + t_{n-1} I \\
A_n &= AB_{n-1}, \frac{1}{n}\operatorname{tr} A_n = -t_n, B_n = A_n + t_n I.
\end{aligned}
\tag{8.18}
$$

We first prove (by induction) that $t_r = b_r$ $(r = 1, 2, \ldots n)$.

By (8.16), we have $t_1 = b_1$.

Assume that $t_i = b_i \quad (i = 2, 3, \ldots k-1)$. (8.19)
From (8.18) we obtain

$$A_2 = AB_1 = A(A_1+t_1I) = A^2 + t_1A$$

$$A_3 = AB_2 = A(A_2+t_2I) = A(A^2+t_1A+t_2I) = A^3 + t_1A^2 + t_2A$$

and, in general,

$$A_k = A^k + t_1A^{k-1} + \ldots + t_{k-1}A, \text{ and by (8.19)}$$

$$= A^k + b_1A^{k-1} + \ldots + b_{k-1}A. \tag{8.20}$$

By (8.17) it follows that

$$\operatorname{tr} A_k = \operatorname{tr} A^k + b_1 \operatorname{tr} A^{k-1} + \ldots + b_{k-1} \operatorname{tr} A. \tag{8.21}$$

But (by the kth equations (8.18))

$$\operatorname{tr} A_k = -kt_k. \tag{8.22}$$

We can rewrite (8.21), using the definition for $S_1, S_2, \ldots$

$$\operatorname{tr} A_k = S_k + b_1S_{k-1} + \ldots + b_{k-1}S_1.$$

$$= -kb_k \quad \text{(by 8.14)} \tag{8.23}$$

Hence from (8.22) and (8.23) we deduce that

$$t_k = b_k \tag{8.24}$$

which is therefore true for all $k(k = 1, 2, \ldots n)$.

We next show that the sequences $\{b_i\}$ and $\{B_i\}$ determine A^{-1}.

From Eq. (8.20), which has been proved valid for $k = 1, 2, \ldots n$, and (8.24) and the last equation of (8.18), we have

$$\begin{aligned} B_n &= A^n + b_1 A^{n-1} + \ldots + b_{n-1}A + b_n I \\ &= 0 \quad \text{(by the Cayley-Hamilton theorem)} \end{aligned}$$

Hence $B_n = A_n + b_n I = 0$, so that

$$A_n = -b_n I. \tag{8.25}$$

But by (8.18)

$$A_n = AB_{n-1}. \tag{8.26}$$

Hence from (8.25) and (8.26), we finally obtain

$$A^{-1} = -\frac{1}{b_n} B_{n-1}. \tag{8.27}$$

Example 8.5
Find (1) the inverse and (2) the characteristic equation of the matrix

$$A = \begin{bmatrix} 1 & 0 & -2 \\ 4 & 1 & 0 \\ 0 & 2 & 1 \end{bmatrix}.$$

Solution
(1) $-b_1 = 3$, so that

$$B_1 = A_1 + b_1 I = \begin{bmatrix} 1 & 0 & -2 \\ 4 & 1 & 0 \\ 0 & 2 & 1 \end{bmatrix} - \begin{bmatrix} 3 & 0 & 0 \\ 0 & 3 & 0 \\ 0 & 0 & 3 \end{bmatrix} = \begin{bmatrix} -2 & 0 & -2 \\ 4 & -2 & 0 \\ 0 & 2 & -2 \end{bmatrix}$$

$$A_2 = AB_1 = \begin{bmatrix} 1 & 0 & -2 \\ 4 & 1 & 0 \\ 0 & 2 & 1 \end{bmatrix} \begin{bmatrix} -2 & 0 & -2 \\ 4 & -2 & 0 \\ 0 & 2 & -2 \end{bmatrix} = \begin{bmatrix} -2 & -4 & 2 \\ -4 & -2 & -8 \\ 8 & -2 & -2 \end{bmatrix}$$

$-2b_2 = -6$, so that $b_2 = 3$

$$B_2 = A_2 + b_2 I = \begin{bmatrix} 1 & -4 & 2 \\ -4 & 1 & -8 \\ 8 & -2 & 1 \end{bmatrix}$$

$$A_3 = AB_2 = \begin{bmatrix} 1 & 0 & -2 \\ 4 & 1 & 0 \\ 0 & 2 & 1 \end{bmatrix} \begin{bmatrix} 1 & -4 & 2 \\ -4 & 1 & -8 \\ 8 & -2 & 1 \end{bmatrix} = \begin{bmatrix} -15 & 0 & 0 \\ 0 & -15 & 0 \\ 0 & 0 & -15 \end{bmatrix}$$

$-3b_3 = -45$, so that $b_3 = 15$.

By (8.27) we have,

$$A^{-1} = -\frac{1}{b_3} B_2 = \frac{1}{15} \begin{bmatrix} -1 & 4 & -2 \\ 4 & -1 & 8 \\ -8 & 2 & -1 \end{bmatrix}.$$

(2) The characteristic equation is

$$\lambda^3 + b_1\lambda^2 + b_2\lambda + b_3 = 0$$

that is,

$$\lambda^3 - 3\lambda^2 + 3\lambda + 15 = 0.$$

8.4 INVERTING A MATRIX WITH COMPLEX ELEMENTS

Although the methods already discussed are suitable for inverting complex matrices (that is, matrices having complex elements), it is sometimes useful to invert such matrices by manipulations of real matrices only.

Various methods which achieve this objective, exist. The one discussed below was refined in [6].

Let $M = A + iB$, have an inverse, say

$$M^{-1} = C + iD$$

then $[A+iB]\,[C+iD] = I$, so that

$$AC - BD = I \tag{8.28}$$

$$BC + AD = 0. \tag{8.29}$$

By pre-multiplying (8.28) and (8.29) by A and B respectively, and adding, we obtain

$$[A^2+B^2]C + [BA-AB] = A. \tag{8.30}$$

By pre-multiplying (8.28) and (8.29) by B and A respectively, and subtracting, we obtain

$$[AB-BA]C + [A^2+B^2]D = -B. \tag{8.31}$$

If A and B commute, then (8.30) and (8.31) become

$$[A^2+B^2]C = A \text{ and } [A^2+B^2]D = -B \text{ respectively.}$$

Hence, so long as $[A^2+B^2]^{-1}$ exists,

$$M^{-1} = C + iD = [A^2+B^2]^{-1}\,[A-iB]. \tag{8.32}$$

Example 8.6
Invert the matrix

$$M = \begin{bmatrix} -1 & 2+i \\ 2+i & -1 \end{bmatrix}.$$

Solution

Here $A = \begin{bmatrix} -1 & 2 \\ 2 & -1 \end{bmatrix}$ and $B = \begin{bmatrix} 0 & 1 \\ 1 & 0 \end{bmatrix}$.

Since A and B commute, we can use (8.32).

$$[A^2+B^2] = \begin{bmatrix} 6 & -4 \\ -4 & 6 \end{bmatrix}, \text{ so that } [A^2+B^2]^{-1} = \frac{1}{10}\begin{bmatrix} 3 & 2 \\ 2 & 3 \end{bmatrix}.$$

It follows that

$$M^{-1} = \frac{1}{10}\begin{bmatrix} 3 & 2 \\ 2 & 3 \end{bmatrix}\begin{bmatrix} -1 & 2-i \\ 2-i & -1 \end{bmatrix} = \frac{1}{10}\begin{bmatrix} 1-2i & 4-3i \\ 4-3i & 1-2i \end{bmatrix}.$$

If A and B do not commute, we have a more complicated situation. On respectively adding and subtracting (8.28) and (8.29) we obtain

$$[A+B]C + [A-B]D = I$$

and

$$[B-A]C + [A+B]D = -I$$

which are written as

$$EC + FD = I$$

$$-FC + ED = -I$$

where $E = A+B$ and $F = A-B$. In matrix notation the above equations are written as

$$\begin{bmatrix} E & F & C \\ -F & E & D \end{bmatrix} = \begin{bmatrix} I \\ -I \end{bmatrix}, \qquad =$$

so that, provided the inverse exists,

$$\begin{bmatrix} C \\ \\ D \end{bmatrix} = \begin{bmatrix} E & F \\ \\ -F & E \end{bmatrix}^{-1} \begin{bmatrix} I \\ \\ -I \end{bmatrix}. \tag{8.33}$$

If $\begin{bmatrix} E & F \\ \\ -F & E \end{bmatrix}^{-1} = \begin{bmatrix} X & Y \\ \\ Z & W \end{bmatrix}$, we obtain the two sets of equations

$$\begin{aligned} EX + FZ = I, \; - FX + EZ = 0 \\ EY + FW = 0, \; - FY + EW = I \end{aligned} \tag{8.34}$$

and

$$\begin{aligned} XE - YF = I, \; ZE - WF = 0 \\ XF + YE = 0, \; ZF + WE = I. \end{aligned} \tag{8.35}$$

Solving (8.34) and (8.35), we find

$$\begin{aligned} X &= F^{-1}E[F+EF^{-1}E]^{-1} = [F+EF^{-1}E]^{-1}EF^{-1} \\ Y &= -E^{-1}F[E+FE^{-1}F]^{-1} = -[F+EF^{-1}E]^{-1} \\ Z &= [F+EF^{-1}E]^{-1} = [E+FE^{-1}F]^{-1}FE^{-1} \\ W &= [E+FE^{-1}F]^{-1}. \end{aligned}$$

It follows from (8.33) and a certain amount of manipulation of the above equations that

$$\begin{aligned} C &= X-Y \\ &= [E+FE^{-1}F]^{-1} \; [FE^{-1}+I] \end{aligned}$$

and

$$\begin{aligned} D &= Z-W \\ &= [E+FE^{-1}F]^{-1} \; [FE^{-1}-I]. \end{aligned}$$

Hence, provided that the various inverses exist,

$$M^{-1} = C + iD = [E+FE^{-1}F]^{-1}\{[FE^{-1}+I] + i[FE^{-1}-I]\}. \quad (8.36)$$

We can also write M^{-1} in a different form.

$$C = X-Y$$

$$= [F+EF^{-1}E]^{-1}\,[EF^{-1}+I]$$

and

$$D = Z-W$$

$$= [F+EF^{-1}E]^{-1}\,[EF^{-1}+I]\,.$$

Hence

$$M^{-1} = [F+EF^{-1}E]^{-1}\,\{[I+EF^{-1}] + i[I-EF^{-1}]\}. \quad (8.37)$$

Example 8.7
Invert the matrix

$$M = \begin{bmatrix} 1 & 1 \\ 1+2i & 1-2i \end{bmatrix}.$$

Solution

$$A = \begin{bmatrix} 1 & 1 \\ 1 & 1 \end{bmatrix},\; B = \begin{bmatrix} 0 & 0 \\ 2 & -2 \end{bmatrix},\; E = \begin{bmatrix} 1 & 1 \\ 3 & -1 \end{bmatrix} \text{ and } F = \begin{bmatrix} 1 & 1 \\ -1 & 3 \end{bmatrix}.$$

Using (8.36), we find

$$M^{-1} = \frac{1}{8}\begin{bmatrix} 1 & 1 \\ 3 & -1 \end{bmatrix}\left(\begin{bmatrix} 2 & 0 \\ 2 & 0 \end{bmatrix} + i\begin{bmatrix} 0 & 0 \\ 2 & -2 \end{bmatrix}\right) = \frac{1}{4}\begin{bmatrix} 2+i & -i \\ 2-i & i \end{bmatrix}.$$

PROBLEMS FOR CHAPTER 8

1) Identify the row operations represented by the following E-matrices.

(i)
$$\begin{bmatrix} 1 & 0 & 0 \\ k & 1 & 0 \\ 0 & 0 & 1 \end{bmatrix}$$

(ii)
$$\begin{bmatrix} 1 & -2 & 0 \\ 0 & 1 & 0 \\ 0 & 0 & 1 \end{bmatrix}$$

(iii)
$$\begin{bmatrix} 1 & 0 & 0 \\ 0 & 1 & 0 \\ 1 & 0 & 1 \end{bmatrix}.$$

2) Find the inverses of the E-matrices in Problem 1 above.

3) Obtain the inverse of the matrix

$$A = \begin{bmatrix} 1 & 0 & 1 \\ 0 & 1 & 0 \\ 3 & 3 & 2 \end{bmatrix}$$

by using appropriate elementary row operations on the partitioned matrix

$$[A \,\vdots\, I]$$

which reduce the A matrix to a unit matrix. The matrix resulting from the application of these E-operations on I is the inverse of A.

4) Make use of the Lagrange interpolation polynomial to find the inverses of the following Vandermonde matrices.

(i)
$$V = \begin{bmatrix} 1 & 1 & 1 \\ -1 & -2 & -3 \\ 1 & 4 & 9 \end{bmatrix}$$

(ii)
$$V = \begin{bmatrix} 1 & 1 & 1 \\ -1 & 1 & -2 \\ 1 & 1 & 4 \end{bmatrix}.$$

5) Given

(i)
$$A = \begin{bmatrix} 0 & 1 & 0 \\ 0 & 0 & 1 \\ -6 & -11 & -6 \end{bmatrix}$$

(ii)
$$A = \begin{bmatrix} 0 & 1 & 0 \\ 0 & 0 & 1 \\ -2 & 1 & 2 \end{bmatrix}$$

find the matrix V such that $V^{-1}AV$ is diagonal.

6) Use Faddeeva's method to invert

(i)
$$A = \begin{bmatrix} 1 & 0 & 1 \\ 0 & 1 & 0 \\ 3 & 3 & 2 \end{bmatrix} \text{ and}$$

(ii)
$$\begin{bmatrix} 0 & 1 & -1 \\ 2 & 2 & 1 \\ 1 & 3 & 1 \end{bmatrix}.$$

In each case write down the characteristic equation of A.

7) Using operations on real numbers only, invert

(i)
$$M = \begin{bmatrix} 1+2i & i \\ -i & 1-i \end{bmatrix}$$

(ii)
$$M = \begin{bmatrix} 1-i & i \\ 1+i & 1 \end{bmatrix}.$$

Solutions to Problems

CHAPTER 1

(1) (a)

$$A + B = \begin{bmatrix} 3 & 1 \\ 1 & 3 \end{bmatrix}, A - 2B = \begin{bmatrix} 0 & -5 \\ -5 & 3 \end{bmatrix},$$

$$AB = \begin{bmatrix} 0 & 4 \\ 5 & -2 \end{bmatrix} \quad BA = \begin{bmatrix} 0 & 5 \\ 4 & -2 \end{bmatrix}.$$

(b) $A + B$, and $A - 2B$ are not defined.

$$AB = \begin{bmatrix} 0 & 0 \\ -1 & 3 \end{bmatrix}, \quad BA = \begin{bmatrix} 4 & 4 & -6 \\ -2 & 0 & 2 \\ 0 & 2 & -1 \end{bmatrix}.$$

(c) $$A + B = \begin{bmatrix} 2 & 2 & -1 \\ -2 & 2 & 0 \\ 2 & 0 & 2 \end{bmatrix}, \quad A - 2B = \begin{bmatrix} -1 & -4 & -1 \\ 4 & -1 & 0 \\ 2 & 0 & -4 \end{bmatrix}$$

$$AB = \begin{bmatrix} 1 & 2 & -2 \\ -2 & 1 & 0 \\ 2 & 4 & 0 \end{bmatrix}, \quad BA = \begin{bmatrix} 1 & 2 & -1 \\ -2 & 1 & 2 \\ 4 & 0 & 0 \end{bmatrix}.$$

(2) $A^2 = I.$

(3)
$$\begin{aligned} PQ &= A\Lambda_1 AA\Lambda_2 A = A\Lambda_1\Lambda_2 A \\ &= A\Lambda_2\Lambda_1 A \quad \text{(since diagonal matrices commute)} \\ &= QP. \end{aligned}$$

(4) (a) $A = \dfrac{1}{2}(A + A') + \dfrac{1}{2}(A - A').$

(see Ex. 1.15)

(b)
$$A = \begin{bmatrix} 1 & 3/2 & 0 \\ 3/2 & 0 & 9/8 \\ 0 & 9/8 & -1 \end{bmatrix} + \begin{bmatrix} 0 & -1 & -2 \\ 1 & 0 & -7/8 \\ 2 & 7/8 & 0 \end{bmatrix}.$$

(5) (i) Hermitian, (ii) skew-Hermitian, (iii) symmetric, (iv) skew-Hermitian, (v) skew-symmetric, (vi) symmetric.

(6)
$$AB = \left[\begin{array}{cc} \begin{bmatrix} 1 & 0 & 0 \\ 0 & 1 & 0 \\ 0 & 0 & 1 \end{bmatrix}\begin{bmatrix} 1 & 0 & 0 \\ 0 & 2 & 0 \\ 0 & 0 & 3 \end{bmatrix} + \begin{bmatrix} 1 \\ 1 \\ 1 \end{bmatrix}[1 \;\; 0 \;\; 0] & \begin{bmatrix} 1 & 0 & 0 & 0 \\ 0 & 1 & 0 & 0 \\ 0 & 0 & 1 & 1 \end{bmatrix} + \begin{bmatrix} 1 \\ 1 \\ 1 \end{bmatrix}[0] \\ [0 \;\; 0 \;\; 0]\begin{bmatrix} 1 & 0 & 0 \\ 0 & 2 & 0 \\ 0 & 0 & 3 \end{bmatrix} + [1]\,[1 \;\; 0 \;\; 0] & [0 \;\; 0 \;\; 0]\begin{bmatrix} 0 \\ 0 \\ 1 \end{bmatrix} + [1]\,[0] \end{array}\right]$$

$$= \begin{bmatrix} \begin{bmatrix} 1 & 0 & 0 \\ 0 & 2 & 0 \\ 0 & 0 & 3 \end{bmatrix} + \begin{bmatrix} 1 & 0 & 0 \\ 1 & 0 & 0 \\ 1 & 0 & 0 \end{bmatrix} \begin{bmatrix} 0 \\ 0 \\ 1 \end{bmatrix} + \begin{bmatrix} 0 \\ 0 \\ 0 \end{bmatrix} \\ \begin{bmatrix} 0 & 0 & 0 \end{bmatrix} + \begin{bmatrix} 1 & 0 & 0 \end{bmatrix} \begin{bmatrix} 0 \end{bmatrix} + \begin{bmatrix} 0 \end{bmatrix} \end{bmatrix}$$

$$= \begin{bmatrix} 2 & 0 & 0 & 0 \\ 1 & 2 & 0 & 0 \\ 1 & 0 & 3 & 1 \\ 1 & 0 & 0 & 0 \end{bmatrix}.$$

(7) $A^3 = 0$, hence order $r = 3$.

(8) If $A^{-1} = \begin{bmatrix} \alpha & \beta \\ \gamma & \delta \end{bmatrix}$, then $\begin{bmatrix} a & b \\ c & d \end{bmatrix} \begin{bmatrix} \alpha & \beta \\ \gamma & \delta \end{bmatrix} = \begin{bmatrix} 1 & 0 \\ 0 & 1 \end{bmatrix}$.

From $a\alpha + b\gamma = 1$, $c\alpha + d\gamma = 0$, we obtain $\alpha = \dfrac{d}{ad - bc}$ and $\gamma = \dfrac{-c}{ad - bc}$.

From $a\beta + b\delta = 0$, $c\beta + d\delta = 1$, we obtain $\beta = \dfrac{-b}{ad - bc}$ and $\delta = \dfrac{a}{ad - bc}$.

(9) (a)

$$\begin{bmatrix} U_1 \\ I_1 \end{bmatrix} = \begin{bmatrix} 1 & Z_1 \\ 0 & 1 \end{bmatrix} \begin{bmatrix} 1 & 0 \\ 1/Z_3 & 1 \end{bmatrix} \begin{bmatrix} 1 & Z_2 \\ 0 & 1 \end{bmatrix} \begin{bmatrix} U_2 \\ I_2 \end{bmatrix}$$

$$= \frac{1}{Z_3} \begin{bmatrix} Z_1 + Z_3 & Z_1Z_3 + Z_1Z_2 + Z_2Z_3 \\ 1 & Z_2 + Z_3 \end{bmatrix} \begin{bmatrix} U_2 \\ I_2 \end{bmatrix}.$$

(b)
$$\begin{bmatrix} U_1 \\ I_1 \end{bmatrix} = \begin{bmatrix} 1 & 0 \\ 1/Z_1 & 1 \end{bmatrix} \begin{bmatrix} 1 & Z_2 \\ 0 & 1 \end{bmatrix} \begin{bmatrix} 1 & 0 \\ 1/Z_3 & 1 \end{bmatrix} \begin{bmatrix} U_2 \\ I_2 \end{bmatrix}$$

$$\frac{1}{Z_1Z_3} \begin{bmatrix} Z_1Z_2 + Z_1Z_3 & Z_1Z_2Z_3 \\ Z_1 + Z_2 + Z_3 & Z_1Z_3 + Z_2Z_3 \end{bmatrix} \begin{bmatrix} U_2 \\ I_2 \end{bmatrix}.$$

(10)
$$\begin{bmatrix} \dot{x}_1 \\ \dot{x}_2 \\ \dot{x}_3 \end{bmatrix} = \begin{bmatrix} 0 & 1 & 0 \\ 0 & 0 & 1 \\ -6 & -11 & -6 \end{bmatrix} \begin{bmatrix} x_1 \\ x_2 \\ x_3 \end{bmatrix} + \begin{bmatrix} 0 \\ 0 \\ 1 \end{bmatrix} [u] .$$

$$y = [1 \quad 0 \quad 0] \begin{bmatrix} x_1 \\ x_2 \\ x_3 \end{bmatrix}$$

(11)(i)

$$X_i^2 = \begin{bmatrix} 1 & 0 \\ 0 & 1 \end{bmatrix} \quad (i = 1, 2, 3).$$

(ii)

$$X_1 X_2 = \frac{1}{4} \begin{bmatrix} i & 0 \\ 0 & -i \end{bmatrix}, \ X_2 X_1 = \frac{1}{4} \begin{bmatrix} -i & 0 \\ 0 & i \end{bmatrix}.$$

Hence $X_1 X_2 + X_2 X_1 = \begin{bmatrix} 0 & 0 \\ 0 & 0 \end{bmatrix}$.

Similarly $X_2 X_3 + X_3 X_2 = X_1 X_3 + X_3 X_1 = \begin{bmatrix} 0 & 0 \\ 0 & 0 \end{bmatrix}$.

It follows that

$$X_i X_j + X_j X_i = \frac{1}{2}\delta_{ij} I \; (i, j = 1, 2, 3) \quad (\delta_{ij} \text{ is the Kronecker delta}).$$

(12)

$$AA' = \begin{bmatrix} 2(p^2 + q^2) & 0 \\ 0 & 2(p^2 + q^2) \end{bmatrix}$$

Hence $p^2 + q^2 = \frac{1}{2}$.

CHAPTER 2

(1) (a) Yes.

(b) Yes, since if $f, g \in G$, then $f(0) = g(0) = 0$ and $(f + g)(0) = 0$ and $\alpha f(0) = 0, \alpha \in R$.

(c) No, since if $f, g \in G$, $f(0) = \alpha = g(0)$, but $(f + g) \in G$ and $(f + g)(0) \neq \alpha$.

(d) Yes.

(e) No.

For example, $f(x) = x^n \in P_n$ (the set of polynomials of degree n)

$$g(x) = -x^n + x \in P_n$$

But $(f + g)(x) = x \notin P_n$.

(f) Yes. Since if $s_1 = (x_1, y_1, z_1)$ and $s_2 = (x_2, y_2, z_2) \in S$

$\mathbf{s} = \mathbf{s}_1 + \mathbf{s}_2 = (x_1 + x_2, y_1 + y_2, z_1 + z_2)$ and

$z_1 + z_2 = (x_1 + x_2) + (y_1 + y_2)$, hence $\mathbf{s} \in S$. Also $\alpha\mathbf{s} \in S$.

(g) No. Let $\mathbf{s}_i = (x_i, y_i, z_i) \in S$ $(i = 1, 2)$.

Then $z_i = x_i + y_i + 1$ $(i = 1, 2)$.

Let $\mathbf{s} = (x, y, z) = \mathbf{s}_1 + \mathbf{s}_2$,

then $z = z_1 + z_2 = (x_1 + x_2) + (y_1 + y_2) + 2$,

hence $\mathbf{s} \notin S$.

(2) (a) (i) $x^2 + 2x + 1$, (ii) $x^2 + 2x + 1$, (iii) $0.x^2 + 0.x + 1 = 1$.

(b) F is not a field (since, for example, 2 has no multiplicative inverse). Hence U is not a vector space.

(3) (i) $(-4, -1, 6)$, (ii) $x^2 + 4x - 2$, (iii) $7 - i$.

(4) $\mathbf{x} = 6\mathbf{x}_1 - 13\mathbf{x}_2 + 4\mathbf{x}_3$.

(5) (a) Dependent. $\mathbf{x}_1$, $\mathbf{x}_2$ and $\mathbf{x}_4$ are linearly independent.

$$\mathbf{x}_3 = \mathbf{x}_1 + 2\mathbf{x}_2 + 0.\mathbf{x}_3.$$

(b) Dependent. $\mathbf{y}_1$ and $\mathbf{y}_2$ are linearly independent

$$\mathbf{y}_3 = 3\mathbf{y}_1 + 2\mathbf{y}_2.$$

(6) P spans a subspace V_1 of R^3. dim $V_1 = 2$. Q spans a subspace V_2 of R^3. Since $\mathbf{y}_3 = \mathbf{y}_1 + 2\mathbf{y}_2$, dim $V_2 = 2$. Let $\mathbf{x} \in V_1$, then $\mathbf{x} = \alpha\,\mathbf{x}_1 + \beta\mathbf{x}_2$ $(\alpha, \beta \in R)$.
But $\mathbf{x}_1 = -\mathbf{y}_1 + \mathbf{y}_2$ and $\mathbf{x}_2 = 2\mathbf{y}_1 - \mathbf{y}_2$.
Hence $\mathbf{x} = (-\alpha + 2\beta)\mathbf{y}_1 + (\alpha - \beta)\,\mathbf{y}_2 \in V_2$. Let $\mathbf{y} \in V_2$, then $\mathbf{y} = a\mathbf{y}_1 + b\mathbf{y}_2$ $(a, b \in R)$.
But $\mathbf{y}_1 = \mathbf{x}_1 + \mathbf{x}_2$ and $\mathbf{y}_2 = 2\mathbf{x}_1 + \mathbf{x}_2$.
Hence $\mathbf{y} = (a + 2b)\mathbf{x}_1 + (a + b)\mathbf{x}_2 \in V_1$. It follows that $V_1 = V_2$.

(7) $[\mathbf{x}]_Q = [-2, 0, 1]'$.

(8) (a)

$$P = \begin{bmatrix} 2 & -1 & 0 \\ -3 & 1 & 0 \\ -1 & 0 & -1 \end{bmatrix}$$

(b) $[\mathbf{x}]_Q = [-2, -3, -1]'$.
$[\mathbf{x}]_S = P[\mathbf{x}]_Q = [-1, 3, 3]'$

(c)

$$P^{-1} = \begin{bmatrix} -1 & -1 & 0 \\ -3 & -2 & 0 \\ 1 & 1 & -1 \end{bmatrix}.$$

CHAPTER 3.

(1) (a) Let $f, g \in P$, then
$D(f + g) = Df + Dg$
$D(\alpha f) = \alpha Df$
(as can be verified).

(b) $\int(f + g) = \int f + \ \ g$
$\int \alpha f = \alpha \int f.$

(2) (a) $D(1) = 0.1 + 0.x + 0.x^2 + 0.x^3$

$D(x) = 1.1 + 0.x + 0.x^2 + 0.x^3$

$D(x^2) = 0.1 + 2.x + 0.x^2 + 0.x^3$

$D(x^3) = 0.1 + 0.x + 3.x^2 + 0.x^3$

$D(x^4) = 0.1 + 0.x + 0.x^2 + 4.x^3$

$$\text{hence } A = \begin{bmatrix} 0 & 1 & 0 & 0 & 0 \\ 0 & 0 & 2 & 0 & 0 \\ 0 & 0 & 0 & 3 & 0 \\ 0 & 0 & 0 & 0 & 4 \end{bmatrix}.$$

(b)

$$A \begin{bmatrix} 1 \\ -2 \\ 1 \\ -1 \\ 2 \end{bmatrix} = \begin{bmatrix} -2 \\ 2 \\ -3 \\ 8 \end{bmatrix} \text{ that is, D (polynomial)} = -2 + 2x - 3x^2 + 8x^3.$$

(3) $I(\mathbf{z}_1) = x \qquad = \frac{1}{2}\mathbf{y}_1 + 1\,\mathbf{y}_2 + 0.\mathbf{y}_3 + 0.\mathbf{y}_4$

$I(\mathbf{z}_2) = x + \frac{1}{2}x^2 \qquad = \frac{1}{2}\mathbf{y}_1 + 1\,\mathbf{y}_2 + \frac{1}{2}\mathbf{y}_3 + 0.\mathbf{y}_4$

$I(\mathbf{z}_3) = x - \frac{1}{2}x^2 + \frac{1}{3}x^3 = \frac{1}{2}\mathbf{y}_1 + 1\,\mathbf{y}_2 - \frac{1}{2}\mathbf{y}_3 + \frac{1}{3}\mathbf{y}_4.$

$$\text{hence } A = \begin{bmatrix} \frac{1}{2} & \frac{1}{2} & \frac{1}{2} \\ 1 & 1 & 1 \\ 0 & \frac{1}{2} & -\frac{1}{2} \\ 0 & 0 & \frac{1}{3} \end{bmatrix}.$$

(4) (i) – (v) are linear transformations, (vi) is not. For example (i).

Let $\mathbf{u} = (x_1, y_1)$ and $\mathbf{w} = (x_2, y_2)$, then

$$f(\mathbf{u} + \mathbf{w}) = f(x_1 + x_2, y_1 + y_2) = (-y_1 - y_2, x_1 + x_2)$$

$$= (-y_1, x_1) + (-y_2, x_2) = f(u) + f(\mathbf{w}).$$

Also $f(k\mathbf{u}) = f(kx_1, ky_1) = (-ky_1, kx_1) = k(-y_1, x_1) = kf(\mathbf{u})$

Since it is the natural basis implied in the question, we have for (i)

$$\mathrm{T}(\mathbf{x}_1) = \mathrm{T}(1, 0) = (0, 1)$$

$$\mathrm{T}(\mathbf{x}_2) = \mathrm{T}(0, 1) = (-1, 0)$$

$$\text{hence } A = \begin{bmatrix} 0 & -1 \\ 1 & 0 \end{bmatrix}.$$

Similarly for (ii)

$$T(\mathbf{x}_1) = \mathrm{T}(1, 0) = (-1, 0)$$

$$T(\mathbf{x}_2) = \mathrm{T}(0, 1) = (0, 1)$$

$$\text{hence } A = \begin{bmatrix} -1 & 0 \\ 0 & 1 \end{bmatrix}.$$

(iii) $A = \begin{bmatrix} 0 & 1 \\ 1 & 0 \end{bmatrix}.$ (iv) $A = \begin{bmatrix} 3 & 0 \\ 0 & 3 \end{bmatrix}.$ (v) $A = \begin{bmatrix} 2 & -1 \\ 3 & 1 \end{bmatrix}.$

(5) (a) $(f \circ g)(x) = f(g(x))$

$$(f \circ g)(\alpha x_1 + \beta x_2) = f[g(\alpha x_1 + \beta x_2)]$$

$$= f[\alpha g(x_1) + \beta g(x_2)] \text{ since } g \text{ is linear.}$$

$$= \alpha f(g(x_1)) + \beta f(g(x_2)) \text{ since } f \text{ is linear.}$$

$$= \alpha(f \circ g)(x_1) + \beta(f \circ g)(x_2)$$

which proves that $f \circ g$ defines a linear transformation.

(b) By Def. 3.5,

$$T_1(\mathbf{e}_1) = a_{11}\mathbf{e}_1 + a_{21}\mathbf{e}_2$$

$$T_1(\mathbf{e}_2) = a_{12}\mathbf{e}_1 + a_{22}\mathbf{e}_2$$

$$T_2(\mathbf{e}_1) = b_{11}\mathbf{e}_1 + b_{21}\mathbf{e}_2$$

$$T_2(\mathbf{e}_2) = b_{12}\mathbf{e}_1 + b_{22}\mathbf{e}_2$$

So that

$$T_1(T_2(\mathbf{e}_1)) = T_1(b_{11}\mathbf{e}_1 + b_{21}\mathbf{e}_2)$$

$$= b_{11}T_1(\mathbf{e}_1) + b_{21}T_1(\mathbf{e}_2)$$

$$= (b_{11}a_{11} + b_{21}a_{12})\mathbf{e}_1 + (b_{11}a_{21} + b_{21}a_{22})\mathbf{e}_2.$$

Similarly

$$T_1(T_2(\mathbf{e}_2)) = (b_{12}a_{11} + b_{22}a_{12})\mathbf{e}_1 + (b_{12}a_{21} + b_{22}a_{22})\mathbf{e}_2.$$

Hence the matrix representing T_1T_2 is

$$\begin{bmatrix} b_{11}a_{11} + b_{21}a_{12} & b_{12}a_{11} + b_{22}a_{12} \\ b_{11}a_{21} + b_{21}a_{22} & b_{12}a_{21} + b_{22}a_{22} \end{bmatrix} = \begin{bmatrix} a_{11} & a_{12} \\ a_{21} & a_{22} \end{bmatrix}\begin{bmatrix} b_{11} & b_{12} \\ b_{21} & b_{22} \end{bmatrix} = AB.$$

(c) From Problem 4, the matrix corresponding to f is

$A = \begin{bmatrix} 0 & -1 \\ 1 & 0 \end{bmatrix}$ and to g, $B = \begin{bmatrix} 0 & 1 \\ 1 & 0 \end{bmatrix}$, hence the matrix

corresponding to $(f \circ g)$ is $\begin{bmatrix} 0 & -1 \\ 1 & 0 \end{bmatrix}\begin{bmatrix} 0 & 1 \\ 1 & 0 \end{bmatrix} = \begin{bmatrix} -1 & 0 \\ 0 & 1 \end{bmatrix}.$

(6) (a) $T(1, 0) = (3, 2)$, $T(0, 1) = (-4, 1)$, hence

$$A_1 = \begin{bmatrix} 3 & -4 \\ 2 & 1 \end{bmatrix}.$$

(b) (i) $T(-1, 1) = (-7, -1)$, $T(2, 1) = (2, 5)$

$$A_2 = \begin{bmatrix} -7 & 2 \\ -1 & 5 \end{bmatrix}$$

or (ii) by Eq. (3.11)

$$A_2 = A_1 P = \begin{bmatrix} 3 & -4 \\ 2 & 1 \end{bmatrix} \begin{bmatrix} -1 & 2 \\ 1 & 1 \end{bmatrix} = \begin{bmatrix} -7 & 2 \\ -1 & 5 \end{bmatrix}.$$

(c) (i) $T(1, 0) = (3, 2) = \frac{5}{2}(1, 1) + \frac{1}{2}(1, -1)$

$T(0, 1) = (-4, 1) = -\frac{3}{2}(1, 1) - \frac{5}{2}(1, -1)$, hence

$$A_3 = \frac{1}{2} \begin{bmatrix} 5 & -3 \\ 1 & -5 \end{bmatrix}$$

or (ii) by Eq. (3.16) $A_3 = H^{-1} A_1$

$$H = \begin{bmatrix} 1 & 1 \\ 1 & -1 \end{bmatrix}, \quad H^{-1} = -\frac{1}{2} \begin{bmatrix} -1 & -1 \\ -1 & 1 \end{bmatrix}$$

hence $H^{-1} A_1 = \frac{1}{2} \begin{bmatrix} 5 & -3 \\ 1 & -5 \end{bmatrix}.$

(d) (i) $T(-1, 1) = (-7, -1) = -4(1, 1) - 3(1, -1)$

$T(2, 1) = (2, 5) = \frac{7}{2}(1, 1) - \frac{3}{2}(1, -1)$, hence

$$A_4 = \frac{1}{2}\begin{bmatrix} -8 & 7 \\ -6 & -3 \end{bmatrix},$$

or (ii) Eq. (3.17) and from (a) and (c) above

$$P^{-1} A_1 P = \frac{1}{2}\begin{bmatrix} 5 & -3 \\ 1 & -5 \end{bmatrix}\begin{bmatrix} -1 & 2 \\ 1 & 1 \end{bmatrix} = \frac{1}{2}\begin{bmatrix} -8 & 7 \\ -6 & -3 \end{bmatrix}.$$

CHAPTER 4

(1) (a) $\ker T = \{[x, y, z] : [x - 2y, 2x - 4y] = [0, 0]\}$

$= \{[2r, r, s], r, s \in R\}$.

Nullity of $T = \dim \ker T = 2$.

(b) $T([x, y, z]) = [x - 2y, 2x - 4y]$

$= [x, 2x] - [2y, 4y]$

$= x[1, 2] - y[2, 4]$.

$S = \{[1, 2], [2, 4]\}$ spans the space Im(T).
Rank of T = dimension of Im(T) = 1.

(c) Also dim Im(T) + dim ker (T) = 1 + 2 = 3.

(2) (i) $\ker T = \{\mathbf{x}: A\mathbf{x} = \mathbf{O}\}$
On solving the system of equation $A\mathbf{x} = \mathbf{O}$, we find

$$\{\mathbf{x}: \mathbf{x} = [0, 0, k]', k \in R\}.$$

Dim ker $T = 1$.

$$\text{Im}(T) = \{y : y = Ax \text{ for all } \mathbf{x} \in R^3\}$$

$=$ space spanned by the linear combination of column vectors of A.

Basis for $\text{Im}(T) = \{[-1, 1]', [2, -1]'\}$.

Dim Im $(T) = 2$.

(ii) $\ker T = \{[2k, k, -k, 0]', [k, 2k, 0, -k], k \in R\}$.

Dim ker $T = 2$.

$\text{Im}(T) = \{[-1, 0, 1]', [-1, 1, 0]'\}$.

Dim Im $(T) = 2$.

(3) $A = (x-2)(x-5)$.

For $x \neq 2$, $r(A) = 3$ (since $|A| \neq 0$).

For $x = 2$, $r(A) = 2$ (since $\mathbf{a}_1 = \mathbf{a}_3$).

For $x = 5$, $r(A) = 2$ (since $\mathbf{a}_3 = \frac{4}{5}(2\mathbf{a}_1 - 3\mathbf{a}_2)$).

(4) $r(A) = 2$.
$\ker T$ has basis $\{[-1, -1, 0, 1]', [-2, -1, 1, 0]'\}$.

We choose P as

$$P = \begin{bmatrix} 1 & 0 & -1 & -2 \\ 0 & 1 & -1 & -1 \\ 0 & 0 & 0 & 1 \\ 0 & 0 & 1 & 0 \end{bmatrix}.$$

With the notation of Theorem 4.5,

$\mathbf{y}_1 = [1, 2, -1, 0]'$ and $\mathbf{y}_2 = [-1, 1, 3, 1]'$ and choose
$\mathbf{y}_3 = [1, 0, 0, 0]'$, $\mathbf{y}_4 = [0, 1, 0, 0]'$.

$$H^{-1} = \begin{bmatrix} 0 & 0 & -1 & 3 \\ 0 & 0 & 0 & 1 \\ 1 & 0 & 1 & -2 \\ 0 & 1 & 2 & -7 \end{bmatrix}.$$

It can now be verified that $H^{-1}AP$ is in the desired form.

(5)

$$[B \vdots AB \vdots A^2B] = \begin{bmatrix} 0 & 1 & 0 \\ 1 & 0 & 3 \\ 0 & -2 & 5 \end{bmatrix}$$

$$\begin{bmatrix} C \\ \hdashline CA \\ \hdashline CA^2 \end{bmatrix} = \begin{bmatrix} 1 & 0 & 1 \\ -1 & -1 & -3 \\ 2 & 5 & 10 \end{bmatrix}.$$

Since the rank of both matrices is 3, the system is both completely controllable and completely observable.

(6) $\Delta = -11$, $\Delta_1 = -11$, $\Delta_2 = 11$, and $\Delta_3 = -22$.
Hence $x_1 = 1$, $x_2 = -1$, $x_3 = 2$.

(7)

$$\text{adj}\, A = \begin{bmatrix} -4 & -3 & 2 \\ 3 & 5 & -7 \\ 5 & 1 & -8 \end{bmatrix}.$$

$$A^{-1} = \frac{-1}{11}\,\text{adj}\, A.$$

(8)

(i) $$\mathbf{x} \times \mathbf{y} = \begin{vmatrix} \mathbf{i} & \mathbf{j} & \mathbf{k} \\ 1 & 2 & -1 \\ 2 & 1 & 3 \end{vmatrix} = 7\mathbf{i} - 5\mathbf{j} - 3\mathbf{k} = (7, -5, -3).$$

(ii) $$(\mathbf{x} \times \mathbf{y}) \times \mathbf{z} = \begin{vmatrix} \mathbf{i} & \mathbf{j} & \mathbf{k} \\ 7 & -5 & -3 \\ 2 & 1 & 2 \end{vmatrix} = -7\mathbf{i} - 20\mathbf{j} + 17\mathbf{k} = (-7, -20, 17).$$

(9) (i) Row (3) – Row (2) $$\begin{vmatrix} 1 & 2 + 3 & (a + 2)(a + 3) \\ 1 & a + 4 & (a + 3)(a + 4) \\ 0 & 1 & 2(a + 4) \end{vmatrix}$$

Row (2) – Row (1) $$\begin{vmatrix} 1 & a + 3 & (a + 2)(a + 3) \\ 0 & 1 & 2(a + 3) \\ 0 & 1 & 2(a + 4) \end{vmatrix}$$

Row (3) – Row (2) $$\begin{vmatrix} 1 & a + 3 & (a + 2)(a + 3) \\ 0 & 1 & 2(a + 3) \\ 0 & 0 & 2 \end{vmatrix}.$$

The result follows on expanding by the first column.

(ii) By the method used in (i), the determinant is reduced to

$$\begin{vmatrix} 1 & a + 2 & (a + 2)^2 \\ 0 & 1 & 2a + 5 \\ 0 & 0 & 2 \end{vmatrix}.$$

The result follows.

(10) Let $\mathbf{x} = [x_1 x_2]'$ and $\mathbf{y} = [y_1 y_2]'$.

$$\text{Then } I + \mathbf{x}\,\mathbf{y}' = \begin{bmatrix} 1 & 0 \\ 0 & 1 \end{bmatrix} + \begin{bmatrix} x_1y_1 & x_1y_2 \\ x_2y_1 & x_2y_2 \end{bmatrix}$$

$$= \left[\begin{bmatrix} 1 \\ 0 \end{bmatrix} + y_1 \begin{bmatrix} x_1 \\ x_2 \end{bmatrix} \begin{bmatrix} 0 \\ 1 \end{bmatrix} + y_2 \begin{bmatrix} x_1 \\ x_2 \end{bmatrix}\right].$$

Hence

$$|I + \mathbf{x}'\,\mathbf{y}| = \begin{vmatrix} 1 & 0 \\ 0 & 1 \end{vmatrix} + y_2 \begin{vmatrix} 1 & x_1 \\ 0 & x_2 \end{vmatrix} + y_1 \begin{vmatrix} x_1 & 0 \\ x_2 & 1 \end{vmatrix} + y_1y_2 \begin{vmatrix} x_1 & x_1 \\ x_2 & x_2 \end{vmatrix}$$

$$= 1 + x_2y_2 + x_1y_1 + 0$$

$$= 1 + \mathbf{x}\,.\,\mathbf{y}$$

CHAPTER 5

(1) The matrix of coefficients is A.

$r(A) = 2.$

The first two columns of A are linearly independent.
Hence $[\alpha, 2, -8]' = a[1, 2, 7]' + b[-2, 1, -4]'$ and $\alpha = -4$.
Choose $z = 5$, and $w = 0$ (arbitrarily), then $x = -3$ and $y = 3$. A particular solution is $[-3, 3, 5, 0]'$. Two linearly independent solutions to $A\mathbf{x} = \mathbf{o}$ are $[7, -4, 0, 5]'$ and $[4, -3, 5, 5]'$.
The general solution is

$$[-3, 3, 5, 0]' + \lambda[7, -4, 0, 5]' + \mu[4, -3, 5, 5]' \quad (\lambda, \mu \in R).$$

(2) (i) $r(A) = 2$, hence there is $3 - 2 = 1$ independent solution.
All solutions are $\lambda[2, -1, 1]'$ $(\lambda \in R)$.

(ii) $r(A) = 2$. Hence 1 independent solution $\lambda[1, 2, 1]'$ $(\lambda \in R)$.

(iii) $r(A) = 3$. Hence $\mathbf{x} = \mathbf{O}$ is the only solution.

(3) (i) $r(A) = 2$ $r(A, \mathrm{b}) = 2$, hence solutions exist.

(ii) $r(A) = 2$ $r(A, \mathrm{b}) = 3$, hence no solution.

(4) Let x_1, x_2 and x_3 be the numbers of fish of each species that can be supported by the food, then

(i)
$$\begin{aligned} 2x_1 + 3x_2 + 2x_3 &= 4650 \\ x_1 + 2x_2 + 3x_3 &= 4350 \\ 2x_1 + x_2 + x_3 &= 2500. \end{aligned}$$

If A is the matrix of coefficients, $r(A) = 3$ (hence unique solution).

$$\text{Since } A^{-1} = \frac{1}{7}\begin{bmatrix} -1 & -1 & 5 \\ 5 & -2 & -4 \\ -3 & 4 & 1 \end{bmatrix},$$

$x_1 = 500, x_2 = 650$, and $x_3 = 850$.

(ii)
$$\begin{aligned} 2x_1 + 3x_2 + 3x_3 &= 4650 \\ x_1 + 2x_2 + x_3 &= 4350 \\ 2x_1 + x_2 + 5x_3 &= 2500. \end{aligned}$$

This time $r(A) = 2$, but $r(A, \mathrm{b}) = 3$, hence no solution exists; that is, in the ratios suggested, not all food could be consumed.

(iii) $r(A) = 2$ and $r(A, \mathrm{b}) = 2$, hence many solutions exist. The general solution is then of the form

$$\mathbf{x} = [2750, -100, 100]' + \lambda[-3, 1, 1]' \ (\lambda \in R).$$

Since $x_i > 0$ $(i = 1, 2, 3)$, $\lambda \in [101, 916]$.

For example, for $\lambda = 750, x_1 = 500, x_2 = 650$, and $x_3 = 850$.

CHAPTER 6

(1) The eigenvalues are $\lambda_1 = 1$ and $\lambda_2 = 2$. The corresponding normalised eigenvectors are

$$\mathbf{x}_1 = \left[\frac{1}{\sqrt{2}}, \frac{1}{\sqrt{2}}\right]' \text{ and } \mathbf{x}_2 = \left[\frac{1}{\sqrt{2}}, -\frac{1}{\sqrt{2}}\right]',$$

hence

$$X = \begin{bmatrix} \frac{1}{\sqrt{2}} & \frac{1}{\sqrt{2}} \\ \frac{1}{\sqrt{2}} & -\frac{1}{\sqrt{2}} \end{bmatrix}.$$

It can now be verified that

(1) $X^{-1}AX = \text{diag}\,\{1, 2\}$

(2) $X'X = I$ so that X is orthogonal.

(2) A is symmetric. $\lambda_1 = 2, \lambda_2 = 7$.
The corresponding normalised eigenvectors are

$$\mathbf{x}_1 = \left[\frac{1}{\sqrt{5}}, -\frac{2}{\sqrt{5}}\right]' \text{ and } \mathbf{x}_2 = \left[\frac{2}{\sqrt{5}}, \frac{1}{\sqrt{5}}\right]'.$$

$$P = \frac{1}{\sqrt{5}}\begin{bmatrix} 1 & 2 \\ -2 & 1 \end{bmatrix} \text{ which is orthogonal, hence}$$

$$P^{-1} = P' = \frac{1}{\sqrt{5}}\begin{bmatrix} 1 & -2 \\ 2 & 1 \end{bmatrix}$$

and $P^{-1}AP = \text{diag}\,\{2, 7\}$.

(3) $|\lambda I - A| = \lambda^3 - 9\lambda^2 + 18\lambda$.

The eigenvalues are $\lambda_1 = 0, \lambda_2 = 3, \lambda_3 = 6$.
Corresponding to λ_1, the normalized vector is $\mathbf{x}_1 = [-2/3, 2/3, 1/3]'$.
Corresponding to λ_2, $\mathbf{x}_2 = [1/3, 2/3, -2/3]'$ and to λ_3, $\mathbf{x}_3 = [2/3, 1/3, 2/3]'$.
Notice that $\mathbf{x}_i \cdot \mathbf{x}_j = \delta_{ij}$.

(4) The eigenvalues are $\lambda_1 = 0$ and $\lambda_2 = 6$. Corresponding to λ_1 the eigenvector is $\mathbf{x}_1 = [(1 + 2i), -1]'$ and to λ_2, the eigenvector is $\mathbf{x}_2 = [1, 1-2i]'$.

$$\tilde{\mathbf{x}}_1 \,.\, \mathbf{x}_2 = [1-2i, -1]' \,.\, [1, 1-2i]' = 0 = \tilde{\mathbf{x}}_2 \,.\, \mathbf{x}_1.$$

(5) The eigenvalues are $+\frac{1}{2}$ and $-\frac{1}{2}$ for all three matrices.

(6) (i) $\lambda^3 - 4\lambda^2 - 4\lambda = 0$

$\text{tr}\, A = 4.$

(ii) $\lambda^3 - 11\,\lambda^2 + 23\lambda - 35 = 0$

$\text{tr}\, A = 11.$

(iii) $\lambda^4 + a_1\lambda^3 + a_2\lambda^2 + a_3\lambda + a_4 = 0$

$\text{tr}\, A = -a_1.$

Notice that the coefficients of the characteristic equation are the same as those in the differential equation (1.1).

(7) (i) $\lambda_1 = 1, \lambda_2 = 4.$

The corresponding normalised column eigenvectors are

$$\mathbf{x}_1 = [\frac{1}{\sqrt{2}}, -\frac{1}{\sqrt{2}}]', \mathbf{x}_2 = [\frac{1}{\sqrt{5}}, \frac{2}{\sqrt{5}}]',$$

and the normalised row eigenvectors are

$$\mathbf{y}_1 = [\frac{2}{\sqrt{5}}, \frac{-1}{\sqrt{5}}], \mathbf{y}_2 = [\frac{1}{\sqrt{2}}, \frac{1}{\sqrt{2}}].$$

Notice that $\mathbf{y}_i \,.\, \mathbf{x}_j = 0 \quad (i \neq j)$

(ii) $\lambda_1 = 5$ and $\lambda_2 = -5.$

The corresponding normalised column eigenvectors are

$$\mathbf{x}_1 = [\frac{2}{\sqrt{5}}, \frac{1}{\sqrt{5}}]' \text{ and } \mathbf{x}_2 = [\frac{1}{\sqrt{5}}, \frac{2}{\sqrt{5}}]',$$

and the normalised row eigenvectors

$$\mathbf{y}_1 = [\frac{2}{\sqrt{5}}, \frac{1}{\sqrt{5}}] \text{ and } \mathbf{y}_2 = [-\frac{1}{\sqrt{5}}, \frac{2}{\sqrt{5}}].$$

Notice the result for $\mathbf{y}_1$ and $\mathbf{y}_2$ is obvious from the symmetry of the matrix A.

Also $\mathbf{x}_i{}' \mathbf{x}_j = 0$ and $\mathbf{y}_i \mathbf{x}_j = 0 \quad (i \neq j)$.

(8) (i)

$$X = \begin{bmatrix} \frac{1}{\sqrt{2}} & \frac{1}{\sqrt{5}} \\ \frac{-1}{\sqrt{2}} & \frac{2}{\sqrt{5}} \end{bmatrix}, \qquad X^{-1} = \frac{\sqrt{2}\sqrt{5}}{3} \begin{bmatrix} \frac{2}{\sqrt{5}} & \frac{-1}{\sqrt{5}} \\ \frac{1}{\sqrt{2}} & \frac{1}{\sqrt{2}} \end{bmatrix}$$

$$Y = \begin{bmatrix} \frac{2}{\sqrt{5}} & \frac{1}{\sqrt{2}} \\ \frac{-1}{\sqrt{5}} & \frac{1}{\sqrt{2}} \end{bmatrix}, \qquad Y^{-1} = \frac{\sqrt{2}\sqrt{5}}{3} \begin{bmatrix} \frac{1}{\sqrt{2}} & \frac{-1}{\sqrt{2}} \\ \frac{1}{\sqrt{5}} & \frac{1}{\sqrt{5}} \end{bmatrix}$$

$$X^{-1}AX = \text{diag}\,\{1, 4\} = Y^{-1}A'Y.$$

(ii)

$$X = \begin{bmatrix} \frac{2}{\sqrt{5}} & \frac{-1}{\sqrt{5}} \\ \frac{1}{\sqrt{5}} & \frac{2}{\sqrt{5}} \end{bmatrix},$$

$X^{-1} = X'$ (X' is the transpose of X)

$X^{-1}AX = \text{diag}\,\{5, -5\}$.

(9) (i) The characteristic equation is

$$(\lambda - 12)(\lambda - 3)^2 = 0.$$

Corresponding to $\lambda_1 = 12$, the eigenvector is $[4, 4, -1]'$. Hence the eigenspace $W_1 = \{[4, 4, -1]'\}$ has dimension 1.
Corresponding to $\lambda_2 = 3$, $W_2 = \{[1, 0, 1]', [0, 1, 1]'\}$, dim $W_2 = 2$.

(ii) The characteristic equation is

$$(\lambda + 1)(\lambda - 1)^2 = 0.$$

Corresponding to $\lambda_1 = -1$, $W_1 = \{[1, -1, 0]'\}$, dim $W_1 = 1$.
Corresponding to $\lambda_2 = 1$, $W_2 = \{[1, 1, 0]', [0, 0, 1]'\}$, dim $W_2 = 2$.

(10)(i) The minimal polynomial $m(\lambda) = \lambda^2 - 15\lambda + 36$.

$$A^4 = 2295\,A - 6804\,I.$$

(ii) $m(\lambda) = \lambda^2 - 1$.

$A^4 = I.$

CHAPTER 7

(1) (i)

$$f(A) = 2A^2 - A + I = \begin{bmatrix} 2 & 3 & 0 \\ 0 & 2 & 0 \\ 0 & 0 & 2 \end{bmatrix}$$

hence $|f(A)| = 8$.
Since the characteristic values of A are 1, 1, and 1, $|f(A)| = [f(1)]^3 = 8$. (See Theorem 7.6).

(ii)

$$f(A) = \begin{bmatrix} 3 & -1 & 0 \\ -1 & 3 & 0 \\ 0 & 0 & 2 \end{bmatrix}$$

$|f(A)| = 16.$

The characteristic values of A are 1, 1, and (-1), $f(1) = 2$, $f(-1) = 4$ hence $|f(A)| = 2.2.4 = 16$.

(2) (i)

$$f(A) = \begin{bmatrix} 5 & -2 & 0 \\ -2 & 5 & 0 \\ 0 & 0 & 3 \end{bmatrix} \text{ which has eigenvalues 3, 3, and 7.}$$

Also as A has eigenvalues 1, 1, (-1), hence (Theorem 7.7) $f(A)$ has eigenvalues $f(1)$, $f(1)$ and $f(-1)$, that is, 3, 3, and 7.

(ii)

$$f(A) = \begin{bmatrix} 6 & 2 & 0 \\ 2 & 6 & 0 \\ 0 & 0 & 8 \end{bmatrix} \text{ which has eigenvalues 8, 8, and 4.}$$

Also $f(1) = 8$ and $f(-1) = 4$.

(3) (i) $C(\lambda) = (\lambda + 1)(\lambda - 1)(\lambda - 2)$.

Since $C(\lambda)$ involves only distinct linear factors, $C(\lambda)$ and $m(\lambda)$ are the same.

(ii) $C(\lambda) = \lambda^3$ and $m(\lambda) = \lambda^2$.

(4) (i) $C(\lambda) = (\lambda - 3)^2(\lambda + 2) = m(\lambda)$.

(ii) No, since the minimal polynomial is not a product of linear factors.

(iii) $W_1 = \ker\{[A + 2I]\}$ and $W_2 = \ker\{[A - 3I]^2\}$.

(iv)

$$[A + 2I] = \begin{bmatrix} 5 & 0 & 0 \\ 1 & 5 & 0 \\ 2 & -1 & 0 \end{bmatrix}, \text{ and } W_1 = \{[0, 0, 1]'\}.$$

$$[A - 3I]^2 = \begin{bmatrix} 0 & 0 & 0 \\ 0 & 0 & 0 \\ -11 & 1 & 25 \end{bmatrix}, \text{ and } W_2 = \{[5, 1, 2]', [0, 5, -1]'\}.$$

(v)

$$P = \begin{bmatrix} 0 & 5 & 0 \\ 0 & 1 & 5 \\ 1 & 2 & -1 \end{bmatrix} \text{ so that } P^{-1} = \frac{1}{25}\begin{bmatrix} -11 & 5 & 25 \\ 5 & 0 & 0 \\ -1 & 5 & 0 \end{bmatrix}.$$

(vi)

$$P^{-1}AP = \begin{bmatrix} -2 & 0 & 0 \\ 0 & 3 & 0 \\ 0 & 1 & 3 \end{bmatrix} = [-2] \oplus \begin{bmatrix} 3 & 0 \\ 1 & 3 \end{bmatrix}.$$

(5) (i) $m(\lambda) = (\lambda + 1)(\lambda - 2)^2$.

(ii) Corresponding to $\lambda + 1, A_1 = [-1]$.

Corresponding to $(\lambda - 2)^2, A_2 = \begin{bmatrix} 2 & 0 \\ 1 & 2 \end{bmatrix}$.

(iii)

$$A_1 \oplus A_2 = \begin{bmatrix} -1 & 0 & 0 \\ 0 & 2 & 0 \\ 0 & 1 & 2 \end{bmatrix}.$$

(iv) $V = W_1 \oplus W_2$.

$W_1 = \ker\{[A + I]\}, W_2 = \ker\{[A - 2I]^2\}$.

$W_1 = \{[1, 0, 3]'\}$.

$W_2 = \{\mathbf{y}, f(A)\mathbf{y}\} = \{[0, 1, 3]', [1, 0\ 2]'\}$ where $f(A) = [A - 2I]$.

Hence $P = \begin{bmatrix} 1 & 0 & 1 \\ 0 & 1 & 0 \\ 3 & 3 & 2 \end{bmatrix}$ so that $P^{-1} = \begin{bmatrix} -1 & 1 & 2 \\ 0 & 2 & 0 \\ -3 & 8 & 4 \end{bmatrix}$

and $P^{-1}AP = A_1 \oplus A_2$.

(6) $m(\lambda) = (\lambda - 2)(\lambda + 1)^2$

$W_1 = \ker\{[A - 2I]\}$, $W_2 = \ker\{[A + I]^2\}$.

$W_1 = \{1, 0, 0]'\}$.

$W_2 = \{\mathbf{y}, f(A)\mathbf{y}\} = \{[1, -1, 1]', [0, 1, 0]'\}$ where $f(A) = [A + I]$.

Hence $P = \begin{bmatrix} 1 & 0 & 1 \\ 0 & 1 & -1 \\ 0 & 1 & 1 \end{bmatrix}$,

(Note the order of the vectors – remember that we require the *upper* Jordan block form)

and $P^{-1} = \begin{bmatrix} 1 & 0 & -1 \\ 0 & 1 & 1 \\ 0 & 0 & 1 \end{bmatrix}$

$$P^{-1}AP = [2] \oplus \begin{bmatrix} -1 & 1 \\ 0 & -1 \end{bmatrix} = \begin{bmatrix} 2 & 0 & 0 \\ 0 & -1 & 1 \\ 0 & 0 & -1 \end{bmatrix}.$$

(7) (i) Since the minimal polynomial for A is $m(\lambda) = \lambda^2 - 1$, $A^2 = I$, hence

$$e^{At} = (1 + \frac{1}{2!}t^2 + \frac{1}{4!}t^4 + \ldots)I + (t + \frac{1}{2!}t^2 + \frac{1}{3!}t^3 + \ldots)A$$
$$= I\cosh t + A\sinh t.$$

(ii) A was shown to be idempotent (see Problem 2, Chapter 1), that is,

$$A^2 = I.$$

(iii) A is nilpotent of order 3 i.e. $A^3 = 0$.

$$e^{At} = I + At + \frac{1}{2!}A^2t^2.$$

(8)

$$\Phi(t) = \begin{bmatrix} 9e^{-2t} - 8e^{-3t} & 6e^{-2t} - 6e^{-3t} \\ -12e^{-2t} + 12e^{-3t} & -8e^{-2t} + 9e^{-3t} \end{bmatrix},$$

$$\int\Phi(t - \tau)B\,u\,d\tau = \begin{bmatrix} 1 - 3e^{-2t} + 2e^{-3t} \\ -1 + 4e^{-2t} - 3e^{-3t} \end{bmatrix}.$$

Hence

$$\mathbf{x}(t) = \begin{bmatrix} 1 + 3e^{-2t} - 4e^{-3t} \\ -1 + 4e^{-2t} + 6e^{-3t} \end{bmatrix}.$$

CHAPTER 8

(1) (i) $R_2 \rightarrow R_2 + kR_1$. (ii) $R_1 \rightarrow R_1 - 2R_2$. (iii) $R_3 \rightarrow R_3 + R_1$.

(2) (i) $\begin{bmatrix} 1 & 0 & 0 \\ -k & 1 & 0 \\ 0 & 0 & 1 \end{bmatrix}$. (ii) $\begin{bmatrix} 1 & 2 & 0 \\ 0 & 1 & 0 \\ 0 & 0 & 1 \end{bmatrix}$. (iii) $\begin{bmatrix} 1 & 0 & 0 \\ 0 & 1 & 0 \\ -1 & 0 & 1 \end{bmatrix}$.

(3)

$$[A \mid I] = \left[\begin{array}{ccc|ccc} 1 & 0 & 1 & 1 & 0 & 0 \\ 0 & 1 & 0 & 0 & 1 & 0 \\ 3 & 3 & 2 & 0 & 0 & 1 \end{array}\right].$$

$R_3 \to R_3 - 3R_2$

$$\left[\begin{array}{ccc|ccc} 1 & 0 & 1 & 1 & 0 & 0 \\ 0 & 1 & 0 & 0 & 1 & 0 \\ 3 & 0 & 2 & 0 & -3 & 1 \end{array}\right].$$

$R_3 \to R_3 - 3R_1$

$$\left[\begin{array}{ccc|ccc} 1 & 0 & 1 & 1 & 0 & 0 \\ 0 & 1 & 0 & 0 & 1 & 0 \\ 0 & 0 & -1 & -3 & -3 & 1 \end{array}\right].$$

$R_1 \to R_1 + R_3$

$$\left[\begin{array}{ccc|ccc} 1 & 0 & 0 & -2 & -3 & 1 \\ 0 & 1 & 0 & 0 & 1 & 0 \\ 0 & 0 & -1 & -3 & -3 & 1 \end{array}\right].$$

$R_3 \to -R_3$

$$\left[\begin{array}{ccc|ccc} 1 & 0 & 0 & -2 & -3 & 1 \\ 0 & 1 & 0 & 0 & 1 & 0 \\ 0 & 0 & 1 & 3 & 3 & -1 \end{array}\right],$$

hence $\begin{bmatrix} -2 & -3 & 1 \\ 0 & 1 & 0 \\ 3 & 3 & -1 \end{bmatrix}$ is the inverse of A.

(4) (i) $L_1(z) = \frac{1}{2}(z^2 + 5z + 6)$

$L_2(z) = -z^2 - 4z - 3$

$L_3(z) = -\frac{1}{2}(z^2 + 3z + 2)$

hence $V^{-1} = \frac{1}{2}\begin{bmatrix} 6 & 5 & 1 \\ -6 & -8 & -2 \\ 2 & 3 & 1 \end{bmatrix}$.

(ii) $L_1(z) = -\frac{1}{2}(z^2 + z - 2)$

$L_2(z) = \frac{1}{6}(z^2 + 3z + 2)$

$L_3(z) = \frac{1}{3}(z^2 - 1)$.

hence $V^{-1} = \frac{1}{6}\begin{bmatrix} 6 & -3 & -3 \\ 2 & 3 & 1 \\ -2 & 0 & 2 \end{bmatrix}$.

(5) (i) $C(\lambda) = \lambda^3 + 6\lambda^2 + 11\lambda + 6 = (\lambda + 1)(\lambda + 2)(\lambda + 3)$

hence $V = \begin{bmatrix} 1 & 1 & 1 \\ -1 & -2 & -3 \\ 1 & 4 & 9 \end{bmatrix}$.

(ii) $C(\lambda) = \lambda^3 + 2\lambda^2 - \lambda - 2 = (\lambda + 1)(\lambda - 1)(\lambda + 2)$

$$\text{hence } V = \begin{bmatrix} 1 & 1 & 1 \\ -1 & 1 & -2 \\ 1 & 1 & 4 \end{bmatrix}$$

(6) (i) $b_1 = -4, b_2 = 2, b_3 = 1$

$$B_1 = \begin{bmatrix} -3 & 0 & 1 \\ 0 & -3 & 0 \\ 3 & 3 & -2 \end{bmatrix}, \qquad A_2 = \begin{bmatrix} 0 & 3 & -1 \\ 0 & -3 & 0 \\ -3 & -3 & -1 \end{bmatrix},$$

$$B_2 = \begin{bmatrix} 2 & 3 & -1 \\ 0 & -1 & 0 \\ -3 & -3 & 1 \end{bmatrix}$$

$$A^{-1} = -B_2$$

$C(\lambda) = \lambda^3 - 4\lambda^2 + 2\lambda + 1.$

(ii) $b_1 = -3, b_2 = -2, b_3 = 5$

$$B_1 = \begin{bmatrix} -3 & 1 & -1 \\ 2 & -1 & 1 \\ 1 & 3 & -2 \end{bmatrix}, \qquad A_2 = \begin{bmatrix} 1 & -4 & 3 \\ -1 & 3 & -2 \\ 4 & 1 & 0 \end{bmatrix},$$

$$B_2 = \begin{bmatrix} -1 & -4 & 3 \\ -1 & 1 & -2 \\ 4 & 1 & -2 \end{bmatrix}$$

$$A^{-1} = -\frac{1}{5}B_2$$

$C(\lambda) = \lambda^3 - 3\lambda^2 - 2\lambda + 5.$

(7) (i)

$$A = \begin{bmatrix} 1 & 0 \\ 0 & 1 \end{bmatrix}, \quad B = \begin{bmatrix} 2 & 1 \\ -1 & -1 \end{bmatrix}, \quad [A^2 + B^2]^{-1} = \frac{1}{5}\begin{bmatrix} 1 & -1 \\ 1 & 4 \end{bmatrix},$$

hence $[A^2 + B^2]^{-1}\,[A - iB] = \frac{1}{5}\begin{bmatrix} 1 - 3i & -1 - 2i \\ 1 + 2i & 4 + 3i \end{bmatrix}.$

(ii)

$$A = \begin{bmatrix} 1 & 0 \\ 1 & 1 \end{bmatrix}, \qquad B = \begin{bmatrix} -1 & 1 \\ 1 & 0 \end{bmatrix}, \qquad E = \begin{bmatrix} 0 & 1 \\ 2 & 1 \end{bmatrix}$$

and $F = \begin{bmatrix} 2 & -1 \\ 0 & 1 \end{bmatrix},$

hence $M^{-1} = \begin{bmatrix} 0 & \frac{1}{4} \\ \frac{1}{4} & \frac{1}{4} \end{bmatrix}\left\{\begin{bmatrix} -1 & 1 \\ 1 & 1 \end{bmatrix} + i\begin{bmatrix} -3 & 1 \\ 1 & -1 \end{bmatrix}\right\}$

$$= \frac{1}{4}\begin{bmatrix} 1 + i & 1 - i \\ -2i & 2 \end{bmatrix}.$$

Bibliography and References

[1] Ayres, F., *Theory and Problems of Matrices,* Schaum.
[2] Barnett, S. and Storey, C., *Matrix Methods in Stability Theory,* Thomas Nelson.
[3] Beaumont, R., A. and Ball R. W., *Introduction to Modern Algebra and Matrix Theory,* Holt, Rinehart and Winston.
[4] Bellman, R., *Introduction to Matrix Analysis,* McGraw-Hill.
[5] Cullen, C., *Matrices and Linear Transformations,* Addison-Wesley.
[6] El-Hawary, M. E., Further Comments on 'A Note on the Inversion of Complex Matrices' *IEEE Trans. on Automatic Control,* April 1975.
[7] Faddeeva, V. N., *Computational Methods of Linear Algebra,* Dover.
[8] Finkbiener, D. T., *Introduction to Matrices and Linear Transformations,* W. H. Freeman.
[9] Gantmacher, F. R., *The Theory of Matrices,* Chelsea N.Y.
[10] Graham, A., *Numerical Analysis,* Transworld Student Library.
[11] Hoffman, K. and Kunze R., *Linear Algebra,* Prentice-Hall.
[12] Kuo, B. C., *Analysis and Synthesis of Sampled – Data Control Systems,* Prentice-Hall.
[13] Lancaster, P. L., *Theory of Matrices,* Academic Press.
[14] Lipschutz, S., *Linear Algebra,* McGraw-Hill.
[15] Martin, D. and Mizel, V. J., *Introduction to Linear Algebra,* McGraw-Hill.
[16] Mirsky, L., *An Introduction to Linear Algebra,* Oxford University Press.
[17] Nering, E. D., *Linear Algebra and Matrix Theory,* John Wiley.
[18] Noble, B., *Applied Linear Algebra,* Prentice-Hall.
[19] Nomizu, K., *Fundamentals of Linear Algebra,* McGraw-Hill.
[20] Paige, L. J. and Swift, J. D., *Elements of Linear Algebra,* Blaisdell.
[21] Pearl, M., *Matrix Theory and Finite Mathematics,* McGraw-Hill.
[22] Pease, M. C., *Methods of Matrix Algebra,* Academic Press.
[23] Porter, B. and Crossley, R., *Modal Control,* Taylor and Francis.
[24] Rosenbrock, H. H. and Storey, C., *Mathematics of Dynamical Systems,* Thomas Nelson.
[25] Tropper, A. M., *Linear Algebra,* Thomas Nelson.

Index

A CATALOG OF SELECTED

DOVER BOOKS

IN SCIENCE AND MATHEMATICS

Astronomy

CHARIOTS FOR APOLLO: The NASA History of Manned Lunar Spacecraft to 1969, Courtney G. Brooks, James M. Grimwood, and Loyd S. Swenson, Jr. This illustrated history by a trio of experts is the definitive reference on the Apollo spacecraft and lunar modules. It traces the vehicles' design, development, and operation in space. More than 100 photographs and illustrations. 576pp. 6 3/4 x 9 1/4. 0-486-46756-2

EXPLORING THE MOON THROUGH BINOCULARS AND SMALL TELESCOPES, Ernest H. Cherrington, Jr. Informative, profusely illustrated guide to locating and identifying craters, rills, seas, mountains, other lunar features. Newly revised and updated with special section of new photos. Over 100 photos and diagrams. 240pp. 8 1/4 x 11. 0-486-24491-1

WHERE NO MAN HAS GONE BEFORE: A History of NASA's Apollo Lunar Expeditions, William David Compton. Introduction by Paul Dickson. This official NASA history traces behind-the-scenes conflicts and cooperation between scientists and engineers. The first half concerns preparations for the Moon landings, and the second half documents the flights that followed Apollo 11. 1989 edition. 432pp. 7 x 10. 0-486-47888-2

APOLLO EXPEDITIONS TO THE MOON: The NASA History, Edited by Edgar M. Cortright. Official NASA publication marks the 40th anniversary of the first lunar landing and features essays by project participants recalling engineering and administrative challenges. Accessible, jargon-free accounts, highlighted by numerous illustrations. 336pp. 8 3/8 x 10 7/8. 0-486-47175-6

ON MARS: Exploration of the Red Planet, 1958-1978--The NASA History, Edward Clinton Ezell and Linda Neuman Ezell. NASA's official history chronicles the start of our explorations of our planetary neighbor. It recounts cooperation among government, industry, and academia, and it features dozens of photos from Viking cameras. 560pp. 6 3/4 x 9 1/4. 0-486-46757-0

ARISTARCHUS OF SAMOS: The Ancient Copernicus, Sir Thomas Heath. Heath's history of astronomy ranges from Homer and Hesiod to Aristarchus and includes quotes from numerous thinkers, compilers, and scholasticists from Thales and Anaximander through Pythagoras, Plato, Aristotle, and Heraclides. 34 figures. 448pp. 5 3/8 x 8 1/2. 0-486-43886-4

AN INTRODUCTION TO CELESTIAL MECHANICS, Forest Ray Moulton. Classic text still unsurpassed in presentation of fundamental principles. Covers rectilinear motion, central forces, problems of two and three bodies, much more. Includes over 200 problems, some with answers. 437pp. 5 3/8 x 8 1/2. 0-486-64687-4

BEYOND THE ATMOSPHERE: Early Years of Space Science, Homer E. Newell. This exciting survey is the work of a top NASA administrator who chronicles technological advances, the relationship of space science to general science, and the space program's social, political, and economic contexts. 528pp. 6 3/4 x 9 1/4. 0-486-47464-X

STAR LORE: Myths, Legends, and Facts, William Tyler Olcott. Captivating retellings of the origins and histories of ancient star groups include Pegasus, Ursa Major, Pleiades, signs of the zodiac, and other constellations. "Classic." – *Sky & Telescope*. 58 illustrations. 544pp. 5 3/8 x 8 1/2. 0-486-43581-4

A COMPLETE MANUAL OF AMATEUR ASTRONOMY: Tools and Techniques for Astronomical Observations, P. Clay Sherrod with Thomas L. Koed. Concise, highly readable book discusses the selection, set-up, and maintenance of a telescope; amateur studies of the sun; lunar topography and occultations; and more. 124 figures. 26 halftones. 37 tables. 335pp. 6 1/2 x 9 1/4. 0-486-42820-6

Browse over 9,000 books at www.doverpublications.com

Chemistry

MOLECULAR COLLISION THEORY, M. S. Child. This high-level monograph offers an analytical treatment of classical scattering by a central force, quantum scattering by a central force, elastic scattering phase shifts, and semi-classical elastic scattering. 1974 edition. 310pp. 5 3/8 x 8 1/2. 0-486-69437-2

HANDBOOK OF COMPUTATIONAL QUANTUM CHEMISTRY, David B. Cook. This comprehensive text provides upper-level undergraduates and graduate students with an accessible introduction to the implementation of quantum ideas in molecular modeling, exploring practical applications alongside theoretical explanations. 1998 edition. 832pp. 5 3/8 x 8 1/2. 0-486-44307-8

RADIOACTIVE SUBSTANCES, Marie Curie. The celebrated scientist's thesis, which directly preceded her 1903 Nobel Prize, discusses establishing atomic character of radioactivity; extraction from pitchblende of polonium and radium; isolation of pure radium chloride; more. 96pp. 5 3/8 x 8 1/2. 0-486-42550-9

CHEMICAL MAGIC, Leonard A. Ford. Classic guide provides intriguing entertainment while elucidating sound scientific principles, with more than 100 unusual stunts: cold fire, dust explosions, a nylon rope trick, a disappearing beaker, much more. 128pp. 5 3/8 x 8 1/2. 0-486-67628-5

ALCHEMY, E. J. Holmyard. Classic study by noted authority covers 2,000 years of alchemical history: religious, mystical overtones; apparatus; signs, symbols, and secret terms; advent of scientific method, much more. Illustrated. 320pp. 5 3/8 x 8 1/2. 0-486-26298-7

CHEMICAL KINETICS AND REACTION DYNAMICS, Paul L. Houston. This text teaches the principles underlying modern chemical kinetics in a clear, direct fashion, using several examples to enhance basic understanding. Solutions to selected problems. 2001 edition. 352pp. 8 3/8 x 11. 0-486-45334-0

PROBLEMS AND SOLUTIONS IN QUANTUM CHEMISTRY AND PHYSICS, Charles S. Johnson and Lee G. Pedersen. Unusually varied problems, with detailed solutions, cover of quantum mechanics, wave mechanics, angular momentum, molecular spectroscopy, scattering theory, more. 280 problems, plus 139 supplementary exercises. 430pp. 6 1/2 x 9 1/4. 0-486-65236-X

ELEMENTS OF CHEMISTRY, Antoine Lavoisier. Monumental classic by the founder of modern chemistry features first explicit statement of law of conservation of matter in chemical change, and more. Facsimile reprint of original (1790) Kerr translation. 539pp. 5 3/8 x 8 1/2. 0-486-64624-6

MAGNETISM AND TRANSITION METAL COMPLEXES, F. E. Mabbs and D. J. Machin. A detailed view of the calculation methods involved in the magnetic properties of transition metal complexes, this volume offers sufficient background for original work in the field. 1973 edition. 240pp. 5 3/8 x 8 1/2. 0-486-46284-6

GENERAL CHEMISTRY, Linus Pauling. Revised third edition of classic first-year text by Nobel laureate. Atomic and molecular structure, quantum mechanics, statistical mechanics, thermodynamics correlated with descriptive chemistry. Problems. 992pp. 5 3/8 x 8 1/2. 0-486-65622-5

ELECTROLYTE SOLUTIONS: Second Revised Edition, R. A. Robinson and R. H. Stokes. Classic text deals primarily with measurement, interpretation of conductance, chemical potential, and diffusion in electrolyte solutions. Detailed theoretical interpretations, plus extensive tables of thermodynamic and transport properties. 1970 edition. 590pp. 5 3/8 x 8 1/2. 0-486-42225-9

Browse over 9,000 books at www.doverpublications.com

Engineering

FUNDAMENTALS OF ASTRODYNAMICS, Roger R. Bate, Donald D. Mueller, and Jerry E. White. Teaching text developed by U.S. Air Force Academy develops the basic two-body and n-body equations of motion; orbit determination; classical orbital elements, coordinate transformations; differential correction; more. 1971 edition. 455pp. 5 3/8 x 8 1/2. 0-486-60061-0

INTRODUCTION TO CONTINUUM MECHANICS FOR ENGINEERS: Revised Edition, Ray M. Bowen. This self-contained text introduces classical continuum models within a modern framework. Its numerous exercises illustrate the governing principles, linearizations, and other approximations that constitute classical continuum models. 2007 edition. 320pp. 6 1/8 x 9 1/4. 0-486-47460-7

ENGINEERING MECHANICS FOR STRUCTURES, Louis L. Bucciarelli. This text explores the mechanics of solids and statics as well as the strength of materials and elasticity theory. Its many design exercises encourage creative initiative and systems thinking. 2009 edition. 320pp. 6 1/8 x 9 1/4. 0-486-46855-0

FEEDBACK CONTROL THEORY, John C. Doyle, Bruce A. Francis and Allen R. Tannenbaum. This excellent introduction to feedback control system design offers a theoretical approach that captures the essential issues and can be applied to a wide range of practical problems. 1992 edition. 224pp. 6 1/2 x 9 1/4. 0-486-46933-6

THE FORCES OF MATTER, Michael Faraday. These lectures by a famous inventor offer an easy-to-understand introduction to the interactions of the universe's physical forces. Six essays explore gravitation, cohesion, chemical affinity, heat, magnetism, and electricity. 1993 edition. 96pp. 5 3/8 x 8 1/2. 0-486-47482-8

DYNAMICS, Lawrence E. Goodman and William H. Warner. Beginning engineering text introduces calculus of vectors, particle motion, dynamics of particle systems and plane rigid bodies, technical applications in plane motions, and more. Exercises and answers in every chapter. 619pp. 5 3/8 x 8 1/2. 0-486-42006-X

ADAPTIVE FILTERING PREDICTION AND CONTROL, Graham C. Goodwin and Kwai Sang Sin. This unified survey focuses on linear discrete-time systems and explores natural extensions to nonlinear systems. It emphasizes discrete-time systems, summarizing theoretical and practical aspects of a large class of adaptive algorithms. 1984 edition. 560pp. 6 1/2 x 9 1/4. 0-486-46932-8

INDUCTANCE CALCULATIONS, Frederick W. Grover. This authoritative reference enables the design of virtually every type of inductor. It features a single simple formula for each type of inductor, together with tables containing essential numerical factors. 1946 edition. 304pp. 5 3/8 x 8 1/2. 0-486-47440-2

THERMODYNAMICS: Foundations and Applications, Elias P. Gyftopoulos and Gian Paolo Beretta. Designed by two MIT professors, this authoritative text discusses basic concepts and applications in detail, emphasizing generality, definitions, and logical consistency. More than 300 solved problems cover realistic energy systems and processes. 800pp. 6 1/8 x 9 1/4. 0-486-43932-1

THE FINITE ELEMENT METHOD: Linear Static and Dynamic Finite Element Analysis, Thomas J. R. Hughes. Text for students without in-depth mathematical training, this text includes a comprehensive presentation and analysis of algorithms of time-dependent phenomena plus beam, plate, and shell theories. Solution guide available upon request. 672pp. 6 1/2 x 9 1/4. 0-486-41181-8

Browse over 9,000 books at www.doverpublications.com

HELICOPTER THEORY, Wayne Johnson. Monumental engineering text covers vertical flight, forward flight, performance, mathematics of rotating systems, rotary wing dynamics and aerodynamics, aeroelasticity, stability and control, stall, noise, and more. 189 illustrations. 1980 edition. 1089pp. 5 5/8 x 8 1/4. 0-486-68230-7

MATHEMATICAL HANDBOOK FOR SCIENTISTS AND ENGINEERS: Definitions, Theorems, and Formulas for Reference and Review, Granino A. Korn and Theresa M. Korn. Convenient access to information from every area of mathematics: Fourier transforms, Z transforms, linear and nonlinear programming, calculus of variations, random-process theory, special functions, combinatorial analysis, game theory, much more. 1152pp. 5 3/8 x 8 1/2. 0-486-41147-8

A HEAT TRANSFER TEXTBOOK: Fourth Edition, John H. Lienhard V and John H. Lienhard IV. This introduction to heat and mass transfer for engineering students features worked examples and end-of-chapter exercises. Worked examples and end-of-chapter exercises appear throughout the book, along with well-drawn, illuminating figures. 768pp. 7 x 9 1/4. 0-486-47931-5

BASIC ELECTRICITY, U.S. Bureau of Naval Personnel. Originally a training course; best nontechnical coverage. Topics include batteries, circuits, conductors, AC and DC, inductance and capacitance, generators, motors, transformers, amplifiers, etc. Many questions with answers. 349 illustrations. 1969 edition. 448pp. 6 1/2 x 9 1/4. 0-486-20973-3

BASIC ELECTRONICS, U.S. Bureau of Naval Personnel. Clear, well-illustrated introduction to electronic equipment covers numerous essential topics: electron tubes, semiconductors, electronic power supplies, tuned circuits, amplifiers, receivers, ranging and navigation systems, computers, antennas, more. 560 illustrations. 567pp. 6 1/2 x 9 1/4. 0-486-21076-6

BASIC WING AND AIRFOIL THEORY, Alan Pope. This self-contained treatment by a pioneer in the study of wind effects covers flow functions, airfoil construction and pressure distribution, finite and monoplane wings, and many other subjects. 1951 edition. 320pp. 5 3/8 x 8 1/2. 0-486-47188-8

SYNTHETIC FUELS, Ronald F. Probstein and R. Edwin Hicks. This unified presentation examines the methods and processes for converting coal, oil, shale, tar sands, and various forms of biomass into liquid, gaseous, and clean solid fuels. 1982 edition. 512pp. 6 1/8 x 9 1/4. 0-486-44977-7

THEORY OF ELASTIC STABILITY, Stephen P. Timoshenko and James M. Gere. Written by world-renowned authorities on mechanics, this classic ranges from theoretical explanations of 2- and 3-D stress and strain to practical applications such as torsion, bending, and thermal stress. 1961 edition. 560pp. 5 3/8 x 8 1/2. 0-486-47207-8

PRINCIPLES OF DIGITAL COMMUNICATION AND CODING, Andrew J. Viterbi and Jim K. Omura. This classic by two digital communications experts is geared toward students of communications theory and to designers of channels, links, terminals, modems, or networks used to transmit and receive digital messages. 1979 edition. 576pp. 6 1/8 x 9 1/4. 0-486-46901-8

LINEAR SYSTEM THEORY: The State Space Approach, Lotfi A. Zadeh and Charles A. Desoer. Written by two pioneers in the field, this exploration of the state space approach focuses on problems of stability and control, plus connections between this approach and classical techniques. 1963 edition. 656pp. 6 1/8 x 9 1/4. 0-486-46663-9

Browse over 9,000 books at www.doverpublications.com

Mathematics–Bestsellers

HANDBOOK OF MATHEMATICAL FUNCTIONS: with Formulas, Graphs, and Mathematical Tables, Edited by Milton Abramowitz and Irene A. Stegun. A classic resource for working with special functions, standard trig, and exponential logarithmic definitions and extensions, it features 29 sets of tables, some to as high as 20 places. 1046pp. 8 x 10 1/2. 0-486-61272-4

ABSTRACT AND CONCRETE CATEGORIES: The Joy of Cats, Jiri Adamek, Horst Herrlich, and George E. Strecker. This up-to-date introductory treatment employs category theory to explore the theory of structures. Its unique approach stresses concrete categories and presents a systematic view of factorization structures. Numerous examples. 1990 edition, updated 2004. 528pp. 6 1/8 x 9 1/4. 0-486-46934-4

MATHEMATICS: Its Content, Methods and Meaning, A. D. Aleksandrov, A. N. Kolmogorov, and M. A. Lavrent'ev. Major survey offers comprehensive, coherent discussions of analytic geometry, algebra, differential equations, calculus of variations, functions of a complex variable, prime numbers, linear and non-Euclidean geometry, topology, functional analysis, more. 1963 edition. 1120pp. 5 3/8 x 8 1/2. 0-486-40916-3

INTRODUCTION TO VECTORS AND TENSORS: Second Edition--Two Volumes Bound as One, Ray M. Bowen and C.-C. Wang. Convenient single-volume compilation of two texts offers both introduction and in-depth survey. Geared toward engineering and science students rather than mathematicians, it focuses on physics and engineering applications. 1976 edition. 560pp. 6 1/2 x 9 1/4. 0-486-46914-X

AN INTRODUCTION TO ORTHOGONAL POLYNOMIALS, Theodore S. Chihara. Concise introduction covers general elementary theory, including the representation theorem and distribution functions, continued fractions and chain sequences, the recurrence formula, special functions, and some specific systems. 1978 edition. 272pp. 5 3/8 x 8 1/2. 0-486-47929-3

ADVANCED MATHEMATICS FOR ENGINEERS AND SCIENTISTS, Paul DuChateau. This primary text and supplemental reference focuses on linear algebra, calculus, and ordinary differential equations. Additional topics include partial differential equations and approximation methods. Includes solved problems. 1992 edition. 400pp. 7 1/2 x 9 1/4. 0-486-47930-7

PARTIAL DIFFERENTIAL EQUATIONS FOR SCIENTISTS AND ENGINEERS, Stanley J. Farlow. Practical text shows how to formulate and solve partial differential equations. Coverage of diffusion-type problems, hyperbolic-type problems, elliptic-type problems, numerical and approximate methods. Solution guide available upon request. 1982 edition. 414pp. 6 1/8 x 9 1/4. 0-486-67620-X

VARIATIONAL PRINCIPLES AND FREE-BOUNDARY PROBLEMS, Avner Friedman. Advanced graduate-level text examines variational methods in partial differential equations and illustrates their applications to free-boundary problems. Features detailed statements of standard theory of elliptic and parabolic operators. 1982 edition. 720pp. 6 1/8 x 9 1/4. 0-486-47853-X

LINEAR ANALYSIS AND REPRESENTATION THEORY, Steven A. Gaal. Unified treatment covers topics from the theory of operators and operator algebras on Hilbert spaces; integration and representation theory for topological groups; and the theory of Lie algebras, Lie groups, and transform groups. 1973 edition. 704pp. 6 1/8 x 9 1/4. 0-486-47851-3

Browse over 9,000 books at www.doverpublications.com

A SURVEY OF INDUSTRIAL MATHEMATICS, Charles R. MacCluer. Students learn how to solve problems they'll encounter in their professional lives with this concise single-volume treatment. It employs MATLAB and other strategies to explore typical industrial problems. 2000 edition. 384pp. 5 3/8 x 8 1/2. 0-486-47702-9

NUMBER SYSTEMS AND THE FOUNDATIONS OF ANALYSIS, Elliott Mendelson. Geared toward undergraduate and beginning graduate students, this study explores natural numbers, integers, rational numbers, real numbers, and complex numbers. Numerous exercises and appendixes supplement the text. 1973 edition. 368pp. 5 3/8 x 8 1/2. 0-486-45792-3

A FIRST LOOK AT NUMERICAL FUNCTIONAL ANALYSIS, W. W. Sawyer. Text by renowned educator shows how problems in numerical analysis lead to concepts of functional analysis. Topics include Banach and Hilbert spaces, contraction mappings, convergence, differentiation and integration, and Euclidean space. 1978 edition. 208pp. 5 3/8 x 8 1/2. 0-486-47882-3

FRACTALS, CHAOS, POWER LAWS: Minutes from an Infinite Paradise, Manfred Schroeder. A fascinating exploration of the connections between chaos theory, physics, biology, and mathematics, this book abounds in award-winning computer graphics, optical illusions, and games that clarify memorable insights into self-similarity. 1992 edition. 448pp. 6 1/8 x 9 1/4. 0-486-47204-3

SET THEORY AND THE CONTINUUM PROBLEM, Raymond M. Smullyan and Melvin Fitting. A lucid, elegant, and complete survey of set theory, this three-part treatment explores axiomatic set theory, the consistency of the continuum hypothesis, and forcing and independence results. 1996 edition. 336pp. 6 x 9. 0-486-47484-4

DYNAMICAL SYSTEMS, Shlomo Sternberg. A pioneer in the field of dynamical systems discusses one-dimensional dynamics, differential equations, random walks, iterated function systems, symbolic dynamics, and Markov chains. Supplementary materials include PowerPoint slides and MATLAB exercises. 2010 edition. 272pp. 6 1/8 x 9 1/4. 0-486-47705-3

ORDINARY DIFFERENTIAL EQUATIONS, Morris Tenenbaum and Harry Pollard. Skillfully organized introductory text examines origin of differential equations, then defines basic terms and outlines general solution of a differential equation. Explores integrating factors; dilution and accretion problems; Laplace Transforms; Newton's Interpolation Formulas, more. 818pp. 5 3/8 x 8 1/2. 0-486-64940-7

MATROID THEORY, D. J. A. Welsh. Text by a noted expert describes standard examples and investigation results, using elementary proofs to develop basic matroid properties before advancing to a more sophisticated treatment. Includes numerous exercises. 1976 edition. 448pp. 5 3/8 x 8 1/2. 0-486-47439-9

THE CONCEPT OF A RIEMANN SURFACE, Hermann Weyl. This classic on the general history of functions combines function theory and geometry, forming the basis of the modern approach to analysis, geometry, and topology. 1955 edition. 208pp. 5 3/8 x 8 1/2. 0-486-47004-0

THE LAPLACE TRANSFORM, David Vernon Widder. This volume focuses on the Laplace and Stieltjes transforms, offering a highly theoretical treatment. Topics include fundamental formulas, the moment problem, monotonic functions, and Tauberian theorems. 1941 edition. 416pp. 5 3/8 x 8 1/2. 0-486-47755-X

Browse over 9,000 books at www.doverpublications.com